DESTROY
BUILD
SECURE

Readings on
Pacification

TYLER WALL
PARASTOU SABERI
WILL JACKSON

Destroy, Build, Secure: Readings on Pacification
Edited by Tyler Wall, Parastou Saberi, Will Jackson

© Red Quill Books Ltd. 2017
Ottawa

www.redquillbooks.com
ISBN 978-1-926958-34-7

Red Quill Books subscribes to a one-book-at-a-time manufacturing process that substantially lessens supply chain waste, reduces greenhouse emissions, and conserves valuable natural resources.

Library and Archives Canada Cataloguing in Publication

Destroy, build, secure : readings on pacification / edited by Tyler Wall, Parastou Saberi, Will Jackson.

Includes bibliographical references.
ISBN 978-1-926958-34-7 (paperback)

1. Social control. 2. Social control—History. 3. Violence—Prevention. I. Wall, Tyler, 1978-, author, editor II. Saberi, Parastou, 1975-, author, editor III. Jackson, Will, 1982-, author, editor

HM661.D48 2016   303.3'3   C2016-907484-6

[RQB is a radical publishing house.
Part of the proceeds from the sale of this book will support student scholarships]

TABLE OF CONTENTS

Introduction 5
    Tyler Wall, Parastou Saberi, Will Jackson

PART 1
PACIFICATION AND THE WAR FOR ACCUMULATION

Fundamentals of Pacification Theory: Twenty-six Articles 13
    Mark Neocleous

Humpty Putting Himself Back Together:
The Hidden Centrality of Participation to Pacification Theory 28
    Stuart Schrader

Counterinsurgency as Global Social Warfare 47
    Ryan Toews

'Going All Out for Shale': Fracking as Primitive Accumulation 68
    Will Jackson, Helen Monk and Joanna Gilmore

PART II
SPACE AND PACIFICATION

Humanizing Domination: On the Role of Urbanism in French Colonial
Pacification Strategies 88
    Parastou Saberi

Fortification and the Fabrication of Colonial Labour 107
    Aaron Henry

Programming the Landscape: Pacification through Landscape Management .. 126
    Philippe Zourgane

Urban Pacification and 'Blitzes' in Contemporary Johannesburg 144
    Christopher McMichael

PART III
PACIFICATION AND POLICING THE SOCIAL ORDER

Pacification Theory: An Empirical Test............................166
    George Rigakos and Aysegul Ergul

Unmanning the Police Manhunt: Vertical Security as Pacification..........188
    Tyler Wall

The War on Purse-Snatchers: Law and Order as Pacification..............208
    Deniz Özçetin

'The Free and Secure University'..................................224
    Biriz Berksoy

Introduction

1. **Know your turf**. Know the people, the topography, economy, history, religion, and culture. Know every village, road, field, population group, tribal leader, and ancient grievance. ... Read the map like a book: study it every night before sleep and redraw it from memory every morning, until you understand its patterns intuitively. Develop a mental model of your area ...[1]

In 2006, in the midst of the escalated insurgency in Iraq, David Kilcullen wrote an influential article in *Military Review*, entitled "'Twenty-eight articles': Fundamentals of company-level counterinsurgency."[2] Kilcullen's 28 on-the-ground recommendations for the American and British forces, which ranged from knowing the ethno-geography of the area and building trust and networks among the local populations (including women), to embarking on small development projects and practicing armed civil affairs, were anchored upon what he conceived to be the two interrelated fundamentals of counterinsurgency. First, counterinsurgency is a combined application of force and politics because insurgency is as much a political threat as it is a military one. Second, counterinsurgency functions through the cooperation and transformation of the local population.[3] A successful counterinsurgency is capable of smooth maneuvering among the liberal boundaries of peace and war, consent and coercion, military and civil spheres, police and military operations.

A few years later, Kilcullen recounted that he intentionally named and organized his 2006 article after T.E. Lawrence's famous "Twenty Seven Articles," published in the *Arab Bulletin* in 1917.[4] Although Lawrence had written his recommendations for the British military to help with the Arab insurgency against the Ottoman Empire, Kilcullen hoped the play with the title would resonate with his military colleagues and "they would readily grasp the allusion."[5] If part of the allusion to T. E. Lawrence was an emphasis on the civilianization of war, the other part was an emphasis on the historical roots of pacification, going beyond the Vietnam War and linking the European colonial wars of the early twentieth century to the 'long war' of the Coalition Forces in Iraq and Afghanistan in the first decade of the twenty-first century.

The history and geography of pacification goes beyond the twentieth-century wars in the Middle East and Asia, however. By the late nineteenth century, Joseph

Gallieni, one of the major figures of French colonial warfare, had already implemented many of Kilcullen's on-the-ground pacification strategies in Indo-China and Madagascar. "An officer who has successfully drawn an exact ethno-graphic map of the territory he commands," Gallieni wrote in 1899, "is close to achieving pacification, soon to be followed by the form of organization he judges most appropriate."[6] His famous *politique des races* (establishing local proxies through whom the colonizer could pacify and rule) is the predecessor of Kilcullen's 'trusted networks' and the emphasis on community partnership and participation in counterinsurgency doctrines. The British soon translated these strategies for conducting imperial policing and small wars across the British Empire.

In their latest works, David Kilcullen[7] and John Nagl[8] have shifted their attention to the necessity of conducting counterinsurgency in the peripherized spaces of urban centers worldwide. Grasping the fact of an increasingly urbanized world, and grounding their analyses in the Malthusian conception of surplus population, the Hobbesian notion of war of all against all, and the Darwinian politics of the survival of the fittest, for Kilcullen and Nagl, future insurgencies will increasingly spring from the slums of the Global South and the *banlieues* of the Global North, rather than from the mountains of Afghanistan. The military, we are told, should shift its focus to urban counterinsurgency and resiliency, learning and appropriating lessons from community policing to urban planning. Despite their claim to novelty, the incorporation of (urban) space in pacification strategies has a long history. Hubert Lyautey, a student of Gallieni, had already perfected the appropriation of urban design and planning as an integral part of colonial pacification in Africa in the late nineteenth and early twentieth centuries.

While the codification of war as peace by the French pacification gurus like Gallieni and Lyautey became the outmost significant reform in the nineteenth century European colonial warfare and domination, the history of pacification goes as far back as the history of modern colonialism and capitalism.[9] Its historical geography stretches from the Americas to Africa and the rest of the globe. The celebrated insights of Gallieni, Lawrence, and Kilcullen appear as translations of the Phillip II 1537 proclamation, when in an attempt to moderate the violence of colonization in the America, he declared:

> They [the pacifiers/conquerors] are to gather information about the various tribes, languages and divisions of the Indians in the province … They are to seek friendship with them through trade and barter, showing them great love and tenderness … By these and other means are the Indians to be pacified and indoctrinated, but in no way are they to be harmed, for all we seek is their welfare and their conversion.[10]

The historical continuity and the geographical diversity in the ideology and practice of pacification are not accidental, particularly when recalling the grand failures of pacification. Rather, they alert us to the importance of pacification to the survival and global expansion of imperialist capitalism throughout the last 400 years. But if the long history of the ideology and practice of pacification direct us to conceive pacification as integral to the historical-materiality of globalized imperialist capitalism today, the concept itself has been largely removed from imperialist military terminology. Instead, counterinsurgency is the prevailing term among contemporary strategists, and pacification, a concept once so prominent in military discourse, has been subsumed under a series of 'more subtle'[11] terms. So why retain a focus on a concept that no longer has the presence it once did?

Beyond the military realm, pacification has appeared in many recent critical treatments of state power, as Christopher McMichael notes in this volume. However, in these critical works, found in a number of disciplines from political science to urban geography, the concept is devoid of any analytical and theoretical dimension. Rather, pacification is often employed simply as a euphemism for 'social control,' 'repression,' or 'policing.' The works in this collection seek instead to contribute to the project of conceiving pacification as integral to the increasingly globalized hegemony of imperialist capitalism. Building upon previous works that have sought to develop pacification as a critical concept,[12] contributors to this volume have aimed to reclaim pacification for critical theory, and in doing so they seek to illustrate the utility of the concept and practice in our attempts to understand imperialist capitalist state power. The chapters in this collection each begin from an understanding that, in Mark Neocleous' terms, there is in the language of pacification an "argument concerning the power of the state in securing the insecurity of capitalist accumulation."[13]

To enable this conceptual development it is vital that we begin from an understanding of the origins of the concept. That pacification is now synonymous with 'counterinsurgency doctrine' is in no small part a result of its importance as the guiding principle of counterinsurgency efforts in Algeria and Vietnam. But, as noted above, pacification, as a concept and practice, has a much longer history dating back to the sixteenth century, and this history, including its point of origin, is important in understanding the inherent links between pacification, colonialism,[14] and capitalism, as well in making sense of the war-police-accumulation nexus.[15]

By positioning pacification as a category of *analysis* (as Schrader explains in this volume), this work has sought to expose, and explore, the dual character of pacification, in which the repression of resistance is but part of the process. Pacification here is employed as a critical concept that helps to understand capitalist/colonialist state power more broadly and its role in the (re)production of bourgeois order. Central to this work has been the exposition of the *productive* nature of pacification that distinguishes it from more cursory uses of the concept in critical literature.

That pacification involves the (re)production of a new social order, as well as the destruction of any opposition, has been clear to the key theorists of pacification from the late sixteenth century onwards, and it is pivotal to the development of a critical concept of pacification in the works collected in this volume. In this sense, pacification involves the production of the multi-dimensional and multi-scale conditions for capital accumulation, and, as Neocleous, Rigakos, and Wall have explained elsewhere, it serves as a "tool for grasping the state-sponsored destruction and reconstruction of social order."[16]

In the first instance, the emphasis on the production of social order makes a key connection with what has historically been understood as the police project.[17] In making this connection explicit, the aim is to expose the means by which the 'docile bodies' and pacified spaces required by state and capital are (re)produced. In doing so, the wider aim is to demonstrate the "conceptual utility of pacification for analyzing the broad vision of police and capital."[18]

From this point, it is crucial to recognize that central to pacification is the logic of security and insecurity. The critical pacification literature seeks to explore the historical and contemporary coupling of bourgeois security and police powers. This critical theory of pacification has its beginnings in the development of an anti-security position[19] that recognizes, at its core, the enduring relevance of Marx's identification of security as the supreme concept of bourgeois society.[20] In pushing the critique of security[21] the aim is to expose how state and capital rely upon the seemingly unquestionable logic of security, and to make clear that critical theory, and any truly radical politics, must "challenge the authoritarian and reactionary nature of security."[22] Security is thus to be understood *as* pacification.[23] The anti-security position, in turn, seeks to understand, and subvert, the hegemony of bourgeois security.

In analyzing the contemporary politics of security through the lens of pacification, the aim is to make sense of the violence employed by state and capital in the (re)production of bourgeois order. Pacification in this sense needs to be understood as a coercion and consent nexus within which violence plays a central role. However, in drawing out the productive function of state/capital violence, the critical theory of pacification avoids reducing the concept to a synonym for repression. Violence is of course as pivotal now as it has ever been in the remaking of capitalist social order, but the works here seek to expose the ever-present iron fist *and* velvet glove of state power. Pacification seeks to break down the apparent division between coercion and consent, uniting them in a single concept, but this project is not without its difficulties. The tension between consent and coercion and the emphasis within different analyses demonstrates that this conceptual development is complex, and the works here are in this regard a collective work in progress.

This present volume developed out of a stream of panels called "Law-Capital-Pacification" organized for the 2014 Critical Legal Studies Conference (Power, Capital, Chaos) at the University of Sussex in Brighton, United Kingdom. Using the larger conference as a way of holding a "workshop" on pacification and capitalist order-building, the conversations proved stimulating and helpful. Hopefully this volume will prove equally stimulating and helpful to those who were not in Brighton, and that conversations and critiques will ensue on the politics of pacification. To help orient readers, we have organized the volume along thematic lines: Part I: Pacification and the War for Accumulation, Part II: Space and Pacification, and Part III: Policing the Social Order. Readers will recognize that there is much substantive and thematic overlap across these sections, so this division of content is merely helpful in organizing what is a diverse but coherent set of chapters.

The first section, Pacification and the War for Accumulation, helps lay the groundwork for understanding projects of pacification as always a war for capital, which is also to say a war against any form of life not subsumed by or convinced of the virtues of accumulation. To speak of pacification, then, is to speak about the ways that accumulation presupposes a certain kind of perpetual war running through the bourgeois order. Now, by a war for accumulation we are thinking primarily of the Marxist notion of the constant, perpetual "social war" that takes as its *sin qua non* the complete subjugation of land, property, and social relations to capitalist dynamics. Hence one idea emerging from this volume and made clear in the first section is that pacification, as a critical concept, is a particularly fruitful analytic to grasp the ways that capital is perpetually on the offensive, always working to 'secure' civil society—'at home' and 'abroad'—through projects of pacification.

Mark Neocleous opens the volume with a punchy chapter playing off of Kilcullen's "Twenty-eight articles." Consisting of his own "twenty-six articles" on the fundamentals of pacification theory, Neocleous provides an important and sharp critique of the strategic logic developed by the high priests of counterinsurgency doctrine such as Kilcullen or Gallieni and Lyautey, who he mockingly calls the "COINdinistas." This chapter sets the stage for much of the chapters that follow by outlining and identifying some of the premises and critiques that are shared across contributions.

Through a discussion of USAID's efforts in Vietnam to garner "participation" among the "local population" in the 1960s, Stuart Schrader's chapter helpfully reminds readers that the destructive aspects of counterinsurgency should not eclipse a more stringent focus on its "reconstructive" efforts. More specifically, Schrader demonstrates that "Pacification has historically required the participation of the pacified," and that efforts of participation-as-pacification required the "creation of 'the people'" or "the community." The next chapter by Ryan Toews forcefully situates counterinsurgency doctrine within a discussion of capitalist imperialism and colonial domination. Confronting the intellectual work of the COINdinistas as the work of imperialist war, Toews ultimately reframes counterinsurgency doctrine as first and foremost a globalizing doctrine of social war.

Will Jackson, Helen Monk, and Joanna Gilmore discuss the politics surrounding the mining practice known as "fracking" through the lens of pacification. Understanding fracking as a form of primitive accumulation via class war, the authors develop their argument by engaging the police responses to the anti-fracking movement in the United Kingdom. This chapter also makes it clear that our formulation of pacification involves more than formal counterinsurgency doctrine or contexts, and instead takes into account "ordinary" state power such as policing, which will become clear in other chapters.

The second section, Space and Pacification, picks up the theme of war and accumulation through a more direct engagement with the spatial dimensions of pacification. These contributions demonstrate the ways that pacification is always already spatial, always already projects of destroying and producing space. In short, the contributions in this section demonstrate the ways that the very real struggles over space, whether understood as territory, land, or landscape, are the concrete, everyday terrain of pacification. Indeed, producing "durably pacified spaces" lies at the heart of projects of producing docile, obedient capitalist subjects.

Parastou Saberi offers a discussion of the role of urbanism in French colonial pacification strategies, specifically of the aforementioned Lyautey and his involvement in Morocco. Saberi demonstrates how spatial strategies were central to colonial warfare, but also how what she calls "humanist urbanism," as a strategy of coding colonial domination as "peace," lay at the heart of Lyautey's approach to pacification. Following this, Aaron Henry engages the spatial strategies of pacification by taking seriously the role of the colonial fort in the eighteenth and nineteenth centuries. Specifically considering the shifting designs and architecture of the fort in regards to making indigenous labor productive, Henry shows how the function of the fort changed from first being a "singular headquarters of power" to "part of a network deployed to develop an extensive spatial field of rule."

Philippe Zourgane discusses the ways that French forces "programmed" the Algerian landscape, meaning how the indigenous landscape was produced in such a way as to foster colonial domination. In the final contribution to this section on space and pacification, Chris McMichael focuses on the police "blitz," or raid, in Johannesburg. McMichael's primary focus is on security policy in the city, and not so much formal counterinsurgency doctrine. Yet what his contribution demonstrates is the ways that the war power and the police power are always spatialized powers, often working in tandem, and carried out through spatial tactics such as the blitz.

The chapters in the third and final section, Pacification and Policing the Social Order, more directly engage pacification through the lens of police power and class war. Here we see policing as the mandate working towards the pacification of those 'unruly' groups and multitudes said to threaten the very foundations of bourgeois social order, such as surplus populations, proletarians, and activists. This includes those tactics and strategies that might be called repressive, yet the central idea animating these contributions is that police power is a productive power to the extent that it works to produce, and reproduce, bourgeois social order.

Hence even police repression, if taken as a sort of negative power that prevents, contains, and eradicates 'threats,' is at once carried out as maneuvers that bring into being a particular historical-material order.

George Rigakos and Aysegul Ergul open the section with an empirical test of pacification theory. To do this, they examine four relevant social trends, namely, the erosion of union membership, the widening gap between rich and poor, the resurgence of political protest, and the rise of employment in public and private police fields. In short, Rigakos and Ergul "treat total policing employment as an empirical barometer of bourgeois insecurity" in relation to income inequality (relative deprivation) and the rise of the "industrial reserve army" via unemployment and deindustrialization. Following this, Tyler Wall provides a discussion on "vertical security as pacification" through an analysis of US domestic police forces adopting aerial drones to police the city. Wall brings together accumulation and pacification by understanding the drone as both security commodity and state weapon designed to essentially hunt the poor.

The final two chapters focus on police power in Turkey as a lens into the political dynamics of pacification. Deniz Özçetin argues that pacification theory provides a useful critique of "law and order" politics in Turkey. She focuses on the ways that the already harsh politics of street crime, specifically the "war on purse-snatchers," has been merged by the security state with the "war on terrorism." Özçetin argues that new legal regulations expanding the discretionary and surveillance powers of police with increased punishments should be understood as nothing less than a strategy of pacification, and a strategy that is aimed primarily at the urban poor. Biriz Berksoy also focuses on pacification strategies in Turkey. Her analysis centers on the politics of the university and Kurdish political and student protest. Berksoy's primary focus is on the ways that during the 2000s campuses became reconfigured through the establishment of what she calls "surveillance-policing networks" that aimed to pacify social opposition to the dominant regimes of accumulation. Berksoy's chapter opens up a larger discussion about the university as a site of pacification under neoliberal regimes of social ordering.

Despite differences in topics, emphasis, and argumentation that exist from chapter to chapter, we hope individual chapters will be read in conversation with other contributions in the volume. In spite of all the messy realities and challenges of any collective work, ultimately, we hope this volume will be seen as a welcomed attempt at collective scholarship that seeks to interrogate the political utility of the concept of pacification for a wider *critique of capital* in an increasingly globalized, urbanized, imperialist capitalist world.

NOTES

1. David Kilcullen, " 'Twenty-eight Articles': Fundamentals of Company-level Counterinsurgency," *Military Review* 86, no. 3 (2006): 103.
2. Ibid.
3. David Kilcullen, *Counterinsurgency* (Oxford: Oxford University Press, 2010), 1-16.
4. Ibid., 24. For E. T. Lawrence's article see, http://wwi.lib.byu.edu/index.php/The_27_Articles_of_T.E._Lawrence
5. Kilcullen, 103.
6. Joseph Gallieni, *Trois Colonnes au Tonkin, 1894-1895* (Paris: 1899), 154.
7. David Kilcullen, *Out of the Mountains: The Coming Age of the Urban Guerrilla* (Oxford: Oxford University Press, 2013).
8. John Nagl, *Knife Fights: A Memoir of Modern War in Theory and Practice* (New York: The Penguin Press, 2014).
9. Mark Neocleous, "Security as Pacification," in *Anti-Security*, eds. Mark Neocleous and George Rigakos (Ottawa: Red Quill Books, 2011), 23-56.
10. Cited in Tzvetan Todorov, *The Conquest of America* (New York: Harper Perennial, 1984), 173.
11. Neocleous, "Security as Pacification," 25.
12. Building on the work done by Mark Neocleous, the beginnings of a critical theory of pacification can be found in Mark Neocleous and George S. Rigakos, eds., *Anti-Security* (Ottawa: Red Quill Books, 2011), and the "On Pacification" issue of *Socialist Studies/Études Socialistes* 9, no. 2 (2013).
13. Mark Neocleous, "A Brighter and Nicer New Life: Security as Pacification," *Social Legal Studies* 20 (2011) 194.
14. See Mark Neocleous, "War as Peace, Peace as Pacification," *Radical Philosophy* 159 (2010): 8-17; Neocleous, *A Brighter and Nicer New Life*,.
15. Mark Neocleous, George Rigakos, and Tyler Wall, "On Pacification: Introduction to the Special Issue," *Socialist Studies/Études Socialistes* 9, no. 2 (2013).
16. Neocleous, Rigakos, and Wall, *"On Pacification,"* 2.
17. Mark Neocleous, *The Fabrication of Social Order* (London: Pluto, 2000).
18. George Rigakos, " 'To Extend the Scope of Productive Labour': Pacification as a Police Project," in *Anti-Security*, eds. Mark Neocleous and George Rigakos (Ottawa: Red Quill Books), 23-56.
19. Neocleous and Rigakos, *Anti-Security*.
20. Karl Marx, "On the Jewish Question" (1844), in *Collected Works, Vol. 3*, Karl Marx and Frederick Engels (London: Lawrence and Wishart, 1975).
21. See Mark Neocleous, *Critique of Security* (Edinburgh: Edinburgh University Press, 2008).
22. Mark Neocleous and George Rigakos, "Anti-Security: A Declaration," in *Anti-Security*, eds. Mark Neocleous and George Rigakos (Ottawa: Red Quill Books), 23-56.
23. Neocleous, "Security as Pacification"; Neocleous, "A Brighter and Nicer New Life*"*.

Part 1
PACIFICATION AND THE WAR FOR ACCUMULATION

Fundamentals of Pacification Theory
Twenty-six Articles

Mark Neocleous

There is today a great deal of talk about counterinsurgency (COIN). Pacification is by no means the same as counterinsurgency, but since their histories are so closely entwined we can do a lot worse than start with that entwinement.

The COINdinistas and security professionals walking the corridors of power and their amateur replicas walking the corridors of universities are never shy of telling us what the world needs to do in the name of security: why we must prevent radicalization, how we should counter insurgency and, more than anything, why we must learn to love capital and the peace that it brings. Those COINdinistas like to communicate in the form of numbered articles, from T. E. Lawrence's "Twenty-seven Articles" of 1917 to David Kilcullen's "Twenty-eight Articles" of 2006. And so, as a contribution to what we might call counter-counterinsurgency – which is a way of taking of sides in the war of accumulation (the world could do with some insurgency in *that* war right now) – offered here in Twenty-six Articles are the fundamentals of pacification theory.

ONE

'Counterinsurgency' has a remarkable history, though because the word is now so commonplace it is a history easily overlooked. According to the *Oxford English Dictionary*, the earliest usage of 'counterinsurgency' appears to be in a US National Security Memoranda issued in November 1961. But then, just three years later, the American state decided that it would instead adopt the concept 'pacification' to describe the counterinsurgency practices being undertaken in Vietnam. Counterinsurgency, we might say, is thus *implicit* in the concept of pacification. Conversely, however, we might also say that *pacification is implicit in the concept of counterinsurgency*. The US Army and Marine Corps *Counterinsurgency Field*

Manual (2006), for example, makes a passing reference to 'pacification' and yet in doing so the extent to which pacification is implicit in the project of counterinsurgency becomes clear. Thus when the ruling class and its political apparatchiks talk of counterinsurgency, we can be sure that they are thinking of pacification.

TWO

Pacification, like counterinsurgency, should not be understood as something that happens 'over there,' so to speak. The National Security Memoranda of November 1961 in which counterinsurgency first appears as an idea is titled "Training for Friendly Police and Armed Forces in Counter-Insurgency, Counter-Subversion, Riot Control and Related Matters."[1] Note the intimate connections made, which tell us something important: in pacification, the police power and the war power are brought together as one. Pacification brings together rebellion and resistance abroad with resistance and rebellion at home. The inside and the outside are understood as a unified order and insurgency-subversion a challenge to that order. The police of specifically named events ('riots') unravels into a whole gamut of other 'related matters,' meaning anything and everything. In pacification there is no 'over there,' because pacification is the systematic colonisation *of the world as a whole* in the name of security and capital.

THREE

To deal with the whole gamut of 'related matters,' pacification requires intelligence. Intelligence is the *sine qua non* of state power, which is why the question of 'intelligence' permeates everything to do with the state: Intelligence Services, Intelligence Staff, Intelligence Committees, diplomatic intelligence, military intelligence, economic intelligence, and on it goes. Pacification is simultaneously a demand *for* intelligence and the generation *of* intelligence (and thus also 'data' of all sorts and now 'big data'), in the name of *raison d'état*. The immense labour of collection and connection of information gathered from here, there and just about anywhere, its recording, reordering, organization and standardization, might be said to be the driving intention of state power. The state is an intelligence-machine and this machine works in the service of pacification.

The rule of 'intelligence' has its origins in the discourse of diplomacy and culture of spying in court society from the fifteenth century onwards: political intelligence was borne of the mutual observation and social competition of court society, overseen by the royal sovereign. Court officials were often understood as 'intelligencers,' which was at that point an alternative name for 'spy,' and this strategic relationship between knowledge and power continues to animate the practices of the modern state. Which is to say: spying is in the nature of the state. The police dream that a few clicks on the computer mouse should be enough to establish who is related to whom, in what degree of intimacy and with what political connection, is the dream of state power.

The state also works on the assumption that the intelligence and data it does have through its overt technologies is not enough, and so engages in a plethora of covert technologies to supplement what it already knows. The combined power of overt and covert intelligence gathering is fundamental to pacification. Pacification requires that the state knows the people, requires organizing for intelligence, and requires that 'intelligence' and 'security' go hand in hand.[2]

FOUR

A great deal of fuss is often made about the secrecy surrounding such intelligence and security practices, but the fact is that to a large degree the state actually wants us to know that such practices take place. A secret is not a thing, but a process of power, and in liberal democracy this process of power operates so that nothing is truly secret. Thus actions described as 'covert' are now far from secret and look much more like 'overt.' A 'covert' operation or drone attack, for example, immediately gets reported in the press with comments from the Head of State. Where assassinations were once kept secret, they are now widely understood as part of the Executive's power. We know enough about the 'covert' assassinations ('targeted killings') organised by the US to know that once a week on a Tuesday the President discusses who is next in line to be assassinated; it is so widely known that it has been dubbed 'Terror Tuesday.' 'Covert' now no longer refers to acts that are kept hidden, but, rather, refers to acts that the Executive knows may be considered highly questionable, possibly illegal and probably immoral, but which it can perform anyway since the body of citizens and their elected representatives can do nothing about it. The power here lies not in keeping the acts secret, but in the process of simulation and dissimulation around the acts.

Simulation and dissimulation are forms of statecraft born at the same time as the rule of intelligence. "Dissimulation is one of the most striking characteristics of our age," Montaigne would observe towards the end of the sixteenth century. As Machiavelli put it in Chapter 18 of *The Prince* earlier that century, the good statesman "must be a great simulator and dissimulator." The comment is a key moment in the development of political theory, as witnessed by the fact that thinker after thinker would repeat the point. "Dissimulation is a great aid," notes Botero in his book on *raison d'état* (1589).[3] The issue is thus not secrecy but how the state simulates and dissimulates concerning its own activities as a state. When it comes to practices of security and intelligence within liberal democracies, the state wants us to know enough about its the 'secret' practices and 'covert' operations so that we behave accordingly. On this view, Wikileaks actually did the state a huge favour by reminding us of just how far the state will go in its endeavours. The state wants us to know that it knows what we know and who we know and how we know them, and it likewise wants us to know exactly what terrors it will carry out in the name of security. Why? Because it wants us to behave accordingly in order that *we internalise our own pacification.*

FIVE

Pacification requires that the state knows the people, what animates them, drives them, and motivates them, not just for the reasons given in Articles Three and Four, but also because the state believes that the enemy *lives among* the people and even that sometimes *the enemy is the people*. The first aim of pacification is to win, retain, or regain the complicity of the pacified. That is, to win, retain, or regain the complicity of the population in the systematic colonisation of the world by capital.

Pacification means shaping the population and building the people so that they govern themselves. (We return to the idea of building in Articles Seventeen to Nineteen.) Shape the governed so that they believe in and accept the mythology which surrounds pacification, such as that 'the police = the people.' Pacification is at its most successful when it is least obvious. This requires mobilizing the population in support of its own pacification.[4] In pacification, the people always remain the prize despite sometimes being the enemy.

SIX

The prize must be fabricated before it can be awarded. After being awarded it must be continually polished (from the root of 'to polish,' which is the same as 'to police') to keep it in the right condition. This is what we understand by the police power. There can be no understanding of pacification without a critical grasp of the police power and a critical grasp of the police power as the fabrication of social order. Police is pacification.

SEVEN

Pacification is war. This war, however, is not a war that would be called 'war' by those specialising in 'war studies' or even 'critical war studies,' for it is a war that concerns the making and remaking of the working class. It is a war for the systematic colonisation of the world and the proletarianization of the people. The struggle over the privatisation of common lands or the length of the working day, to give just two examples, are thus part of a protracted war between the capitalist class and the working class.

This is a war that is conducted by the bourgeoisie under the guise of 'peace and security' and 'law and order.' If "peace itself is a coded war," as Foucault suggests (and as Foucauldians like to repeat time and again though without telling us very much about what all the fighting is actually about), we also insist that the war in question is nothing less than the most significant war in human history. In other words, pacification is class war. Pacification is also therefore liberal peace.

EIGHT

'Peace' derives from the Middle English and Old French *pais*, itself adapted from the Latin *pacem,* which was in turn derived from *pax*. In the Roman tradition, *pax* has as much affinity with the word 'dominance' as it does 'peace.' A declaration of 'peace' as the constitution of the *pax* (in the sense of *Pax Romana*, *Pax Britannica*, *Pax Americana*) implies not an agreement between equals but the unconditional surrender of the conquered, who are thereby 'pacified.' "The Romanes used to say, that their Generall had Pacified such a Province, that is to say, in English, Conquered it," comments Hobbes.[5] This is why we speak of 'imposing the peace' and 'dictating the peace.' What is connoted by *pax* and thus 'peace' is not an absence of conflict or making of a pact but, rather, the imposition of hegemony and domination achieved through conquest and maintained by arms. *Pax*/peace was and is a victor's peace—"the Countrey was *Pacified* by Victory," as Hobbes puts it—achieved through violence. The *pax*/peace is pacification in action. For pacification to be successful, war must be believed to be 'peace' – liberal peace, codified in law.

NINE

The violence of pacification is never against the law, outside the law, or exceptional to law. After 15 years of service to the British state in that state's pacification of Kenya, Cyprus, Malaya, and Oman, Frank Kitson could write that "no country which relies on the law of the land to regulate the lives of its citizens can afford to see that law flouted by its own government, even in an insurgency situation. In other words, everything done by a government and its agents in combating insurgency must be legal."[6] One can see this from the ways in which Kitson, a man of action admired by the military and the political right, was also well-liked by liberals, for liberalism likes nothing better than for pacification to be exercised with liberal values, most notably in the belief that even the most belligerent COINdinistas believe in abiding by the law. To succeed, pacification must be legal as well as liberal.

TEN

The way Kitson continues is telling, however, and it is telling because it captures something fundamental about pacification and law. "This does not mean that the government must work within exactly the same set of laws during an insurgency as existed beforehand," he says. "It is a function of government to make new laws when necessary," and so to say that everything that is done must be legal "does not even mean that the law must be administered in exactly the same way during an uprising as it was in more peaceful times." Rather, "a government has the power to modify the way in which the law is administered if necessary, for the wellbeing of the people. … It is therefore perfectly normal for governments not only to introduce Emergency Regulations as an insurgency progresses, but also to counter advantages which insurgents may derive from, for example, the intimidation of juries and witnesses, by altering the way in which the law is administered." This is a classic reformulation of what we know about law: law follows the police power obediently and carries the bags of the war power subserviently. Pacification enacts changes in the law to make these powers possible, acceptable and legal.

ELEVEN

Pacification is not a state of exception and does not require a state of exception. Excited by the idea that they might be living in exceptional times, many scholars and activists have recently also been taken up by the idea that this is some kind of 'state of exception.' It is not. 'Exception' is indeed central to the state discourse of pacification, in that to think that we are in 'exceptional times' is exactly what the state wants us to think. But pacification is not an exceptional mode of governing and there is nothing exceptional about pacification. The police wars at the heart of pacification are the norm, not an exception to the norm.

Pacification does, however, love the language of 'emergency,' as Kitson's comments show and as a cursory glance at state practice reveals (Emergency Regulations, Emergency Measures, Emergency Laws, Emergency Planning, Emergency Controls, and on it goes), but this is because of what emergency allows and should in no way be taken to mean the same as exception. Note Kitson's comment, "it is perfectly normal." In bourgeois order, pacification is normal; for pacification, bourgeois order is normal. For pacification to be truly successful, the people must believe that the state and capital are normal.

What is 'normal' is the problem, not what is called 'exceptional.' And what is normal today is capital.

TWELVE

Pacification requires a series of interlocking measures, ranging across what scholars like to call 'the military,' 'the police,' 'the economy,' 'the cultural field,' the 'intelligence services' and through the whole gamut of security technologies. In pacification, these things are never clearly distinguishable. The state is a unity, despite that unity appearing to be a complex of organizations and institutions that *appear* to the population to be distinct and sometimes even at odds with each other; this unity is fundamental to pacification.

Yet so too is obscuring that unity: what looks like disunity or even disorganisation is, from the perspective of pacification theory, simply a continuation of a fundamental principle of the police power. Almost everything in pacification is therefore 'inter-agency,' from the collection of intelligence to the collection of the trash. Pacification is therefore never the task of one institution. Understanding pacification requires understanding the state and a grasp of the social totality.

THIRTEEN

These interlocking political, economic, military and police measures are held together by the broad concepts under which pacification operates: 'peace and security,' 'law and order.' With this in mind we can now further refine the points made in Articles Two, Three and Ten.

FOURTEEN

In pacification, the war power is police power; the ideology of security holds them together.

FIFTEEN

In pacification, the police power is war power; the ideology of security holds them together.

SIXTEEN

To bring the war power in line with the police power is by no means to succumb to either of the two myths of contemporary thought: that we are witnessing the 'policization of war' and the 'militarization of police.'

The 'militarization of police' thesis presupposes that there is some kind of 'better' policing, less obviously 'militarized' perhaps. What is usually pointed to here is something along the lines of 'community policing.' Yet any serious analysis shows that community policing is no less problematic and no less part of the war power than a water cannon. There is a good reason why counterinsurgency theorists from General Galliéni and Colonel Lyautey through to Frank Kitson and then on to the recently published US Army and Marine Corps *Counterinsurgency Manual* never stop telling us how important community policing is. And why do they do so? Because if any instance of the police power could be described as core to capital's war of accumulation, it is what passes as community policing. Which is why community policing is one of the core liberal myths and at the heart of pacification. Summing up 50 years of counterinsurgency research up until 2006, one RAND document produced for the Office of the Secretary of Defense commented that "pacification is best thought of as a massively enhanced version of the 'community policing' technique that emerged in the 1970s." Going on to say that community policing is centered on the concept of building "bonds of trust" with local residents, businesses, groups and churches, the document claims that "pacification is simply an expansion of this concept to include greater development and security assistance."[7]

SEVENTEEN

Building bonds of trust to achieve greater development and security? The terminology in that RAND document is important. Understanding why takes us to the core of pacification, and picks up on a point made in Article Five.

When the US embarked on its pacification of Vietnam, it featured a conceptual triad: Build-Secure-Clear. This was a reversal of the French mode of pacification in Indochina and elsewhere: Clear-Secure-Build. Apropos of Articles Thirteen to Fifteen, note the references to security in both triads. But note as well the references to 'Build.' 'To Build' features as pacification's core because it is founded on the concept of improvement at the heart of bourgeois ideology: the bourgeoisie has since its rise to power sought to *build a world after its own image*, and the ideology of improvement is central to this project.

By the eighteenth century 'improvement' had come to permeate political thought, political economy and property law. In its original meaning it refers to "the turning of a thing to profit or good account; making the most of a thing for one's own profit." On this basis the idea of 'improvement' would play a central role in liberal thinking and the ideological development of the bourgeois class, not least in the 'war on waste' it has waged for five hundred years (war on wasted time, war on wasted labour, war on wasted land). Agriculture can always be improved by enclosure; commerce and industry can always be improved with better transport, communications, and the division of labour; labour can always be improved with better discipline, less welfare, and tougher prisons; transport can always be improved with better roads, canals, and bridges; towns and cities can always be improved with better streets and lighting. For the bourgeoisie, *the world must be improved* and it must be improved by coming under the rule of the bourgeoisie.

On the one side, the ideology of improvement is at the heart of class war. Picking up on the bourgeois desire to cultivate the self—moral well-being and economic profit combined and achieved through self-discipline (what today our Masters now insist we call 'resilience'; see Article Twenty-six)—improvement is also the basis of the idea of cultivating the world. 'Improvement' came to be applied to the enclosures and to the making and remaking of the working class; it was thus crucial to the early class war that created and disciplined the proletariat.

It also came to permeate the justifications made in international law for the appropriation of land and resources across the globe.

On the other side, because the idea of improvement makes fundamental assumptions about the *proper* use of nature and thus nature as *property*, what counts as the *proper* use of one's own person (one's mind, body, and labour as *property*) and one's time and thus *propriety*, and thus what counts as a *proper order*, it is therefore also very much a question of police power. Thus we find that in Britain during the eighteenth century a number of Improvement Acts were passed and several hundred new statutory bodies were formed under the rubric of Improvement Commissions, but the Improvement Acts were also known as 'police acts' and the Improvement Commissions were known as 'police commissions.' So common was the idea that the Improvement Acts were police acts that they were often understood as the 'Improvement Act police.'

These two sides remind us of how easily the war power and the police power coincide as the bourgeoisie seeks to improve and *build* the world anew, again and again. There can be no pacification theory without grasping this basic fact. That is, there can be no pacification theory without grasping that pacification has for 500 years been a project not of crushing resistance or breaking rebellion (though it certainly does involve those things) but of *building bourgeois order*. Pacification is a construction project.

EIGHTEEN

Pacification is therefore a tricky business: secure and build, build and secure. This is the core tension in the whole of bourgeois order as Marx and Engels point out in the *Manifesto*. The constant revolutionising of the instruments and relations of production and with them the established relations of society that comes with the impetus to keep building and rebuilding bourgeois order means that the same order is one of everlasting transformation and insecurity. When all that is solid melts into air again and again, we are left with nothing but our insecurities. This tension between the *insecurity* that permeates the whole of the social order as it gets transformed over and over and the idea that *security* is the supreme concept of this very order lies at the heart of bourgeois modernity. Managing that tension is at the heart of pacification.

NINETEEN

"To build, this is the goal, and the unique goal, of every colonial war." So said Lieutenant Colonel Lyautey when leading the French colonising process. He was extending the pacification doctrine of his predecessor, General Gallieni. "The best means for achieving pacification in our new colony is provided by combined application of force and politics," said Gallieni. By 'force' and 'politics' he means 'destruction' and 'reconstruction' respectively. Gallieni writes:

> We should turn to destruction only as a last resort and only as a preliminary to better reconstruction. We must always treat the country and its inhabitants with consideration, since the former is destined to receive our future colonial enterprises and the latter will be our main agents and collaborators in the development of our enterprises. Every time that the necessities of war force one of our colonial officers to take action against a village or an inhabited center, his first concern, once submission of the inhabitants has been achieved, should be reconstruction of the village, creation of a market, and the establishment of a school.

This is the reason why whenever one opens a pacification text one finds the idea that violence is always presented as part and parcel of a *productive* or *constructive* or *reconstructive* application of politics in general. For Gallieni, destruction and reconstruction go hand-in-hand. Yet although "it is by combined use of politics and force that pacification of a country and its future organization will be achieved," it is "political action [that] is by far the more important," deriving as it does "from the organization of the country and its inhabitants." Thus it is the politics/reconstruction that counts, because this is where real pacification lies. "A country is not conquered and pacified when a military operation has decimated and terrorized its people. Once the initial shock passes, a spirit of revolt will arise among the masses, fanned by a feeling of resentment which has been created by the application of brute force." Rather, one can only speak of successful pacification when "the country becomes more civilized, markets are reopened, trade is reestablished. It is necessary, on the one hand, to study and satisfy the social requirements of the subject people and, on the other hand, to promote the development of colonization, which will utilize the natural resources of the soil and open the outlets for European trade."

Gallieni's suggestion is a commonplace that links the practice of pacification across five hundred years of bourgeois modernity. As Lyautey puts it, pacification takes place "not as matter of destroying [people], but of transforming." Why transforming? Because the purpose of pacification is the fabrication of social order. And 50 years later, when the French were rethinking their role on the world stage, this was still the case. In 1957 at Supreme Headquarters Allied Powers Europe (SHAPE), one of NATO's two strategic commands, General Allard of France commented in a speech that the war against the various communist and socialist movements then in existence had to combine a host of social and economic programs as well as 'pure' military action. "I shall classify these various missions under two categories: Destruction and Construction. These two terms are inseparable." He then goes on to expand on these terms. The meaning of destruction is fairly clear: the co-ordinated activity of army and associated state powers to "chase and annihilate" and "deal spectacular blows." Construction, however, means "building the peace," "organizing the people," persuading the people "by the use of education" and, ultimately "preparing the establishment of a new order." As he puts it, "to destroy without building up would mean useless labor; to build without first destroying would be a delusion"—*"this is the task of pacification."*[8]

We have then both the historical basis and the conceptual foundation of the French triad Clear-Secure-Build. The very same principle is in the American rethinking of the triad in the 1960s—Build-Secure-Clear—which is why it coincides with Development Theory of the time. The assumption built into that theory is that infrastructure improvements would facilitate trade and commerce and thereby prevent rebellion.

Pacification is a productive power to build an international order in which capital can thrive and thus a social order founded on the exploitation of wage labour. Building markets is the process of systematic colonisation. *Nation building*, as the Americans liked to call it, but nation building through the building of markets and thus, in effect, *building the market system itself*. Clear-Secure-Build or Build-Secure-Clear: it does not matter to us what the difference is, for we are interested in the consistency of the logic. And what do we find in the US *Counterinsurgency Manual*? The idea that Clear-Hold-Build is a core principle of counterinsurgency.[9]

We might note in passing that 'Security' appears to have disappeared from the last of these triads, but as the *Manual* makes clear in the pages that follow, to Hold really means to Secure.[10] The issue to focus on here, however, is the sustained presence of Build.[11] The *Counterinsurgency Manual* then explicates this in terms of classical police theory, from clearing the trash to managing the market: build houses, build schools, build clinics, build roads, build bridges (literal and metaphorical), build culture, build support, build morale, build the trust and confidence of the people, and above all *build the market*. Pacification is the building of bourgeois order, the very core of what we understand as the constitution of capital and the fabrication of wage labour. This is because at the heart of pacification is the need to *build for accumulation*. "Money is ammunition."[12]

TWENTY

To add to Articles Two, Three, Seventeen, Eighteen, and Nineteen, we are now in a position to state that Pacification = War-Police-Accumulation.

TWENTY-ONE

Pacification requires networks of welfare. The COINdinistas are right: pacification is armed social work.[13] From a critical perspective we have long been aware of how central social work is to the production and reproduction of wage labour and social order: the welfare state is still a form of state and thus has to be read through the lens of class power; this is the police power as social work and social work as police power. The techniques of political administration thereby become central to pacification. The 'people's welfare' is a form of police ideology and opens the space for interventions into the population, enabling the *welfare* state, as state power, to be part and parcel of the pacification process. This is why it is called 'social security': it is a project for securing the social.

TWENTY-TWO

Pacification therefore seeks to build trust and build networks; it seeks to build trusted networks. This is the true meaning of phrases such as 'hearts and minds' (hearts feel the trust; minds believe the trust is real) and 'peaceful penetration' (a term adopted by Lyautey from French colonial traders who sought to build trust in the market in the mind of the colonised).[14] Trust spreads in communities and populations as the market penetrates our being. Trust must ultimately be a trust in the state and the security it offers on the one hand, and in capital and the opportunities it affords on the other. Once the population trusts the state and trusts the ruling class, pacification will be as complete as it can ever be.

TWENTY-THREE

Pacification is never complete; it is a permanent process.

TWENTY-FOUR

Security used to be closely and sometimes solely allied with the idea that the state provides protection, but we are told now that security and protection are a 'shared responsibility' which requires the preparation of the people. This is the basis of the new 'whole of government' approach to security that has emerged, where the 'whole' includes 'the people.' As recently articulated, this logic has been to inform and empower a broader range of people and institutions to become a part of the architecture of security. This requires a number of things. First and foremost, to become part of the architecture of security citizens have to become gatherers of intelligence. This is the basis of the proliferation of techniques such as "Suspicious Activity Reporting Initiatives," "National Terrorism Advisory Systems," and the command "If You See Something, Say Something."[15] Because the threat to security and order can come from anywhere and anyone, so we are instructed that we must all keep watch. Citizens should act on anything suspicious *even when they think it might not be anything*. This is the police logic par excellence: anything slightly out of order is by definition suspicious and thus a security threat. The intelligence needed is intelligence on the population, possessed by the population, *and to be gathered by the population itself on itself.* If pacification is at its most successful when the people are mobilised in their own pacification and depends on intelligence, then we may add that in the name of security citizens are to themselves become the 'intelligencers of the modern world.' Pacification now demands that we must all spy on one other.

TWENTY-FIVE

What this also means, however, is that 'if you see something, say something' could be turned around and offered to us in a slightly revised form: 'if you *do* something, someone will be watching you.' Your neighbours, colleagues, family, friends, lovers, will be watching you, just as you will be watching them. We are reaching the intensified highpoint of a world in which security is the supreme concept, making every human subject a suspicious person.

The phrase 'suspicious person' has a double-meaning. 'I am a suspicious person' connotes being inclined to suspect: I am a person who has suspicions. But 'I am a suspicious person' also connotes giving grounds for suspicion: I am a person who is suspected by others. The suspicious mind is policed by the person with a suspicious mind, and we are all in both categories. The intensified highpoint of a world in which security is the supreme concept is thus a world in which we are rendered unable to engage in one of the most fundamental acts of humanity. "Winston had never been able to feel sure … whether O'Brien was a friend or an enemy."[16] Pacification means that that we play a doubly suspicious role as security operative and potential terrorist, amateur detective and budding conspirator. The only possible outcome is the debasing of any attempt at human solidarity. Pacification is thus not only exploitation, it is also alienation.

TWENTY-SIX

'Becoming part of the architecture of security' is premised on a doctrine of pre-emption and preparation, a technology of power that requires us to be always already 'ready.' Mobilisation appears now in the lingua franca of deterrence, prevention, pre-emption. The population must believe that deterring the enemy, preventing attacks, and pre-empting disaster are natural, normal, and necessary. This is where the question of resilience comes to the fore.

'Resilience' has in the last decade become one of the key political categories of our time, and one reason it has done so is because its key connotation is the capacity to recover from a shock or crisis. In this new discourse, 'bouncebackability' is the defining motif. The point of this preparation and resilience and this preparation for resilience is to nurture and improve the citizen-subjects in *readiness* for the coming disasters/attacks (hence the plethora of projects organised along the lines of 'ready.gov'). It is the people who are to be readied and prepared as part of the new security architecture. If the nation is an imagined community, it is now to be imagined as a community resilient and prepared. If 'Don't Be Scared, Be Prepared' is the official motto of our times, then 'Be Prepared, Be Pacified' is the unofficial. Pacification is now the preparation of the people for their own pacification.

NOTES

1. *National Security Memorandum 114*, November 22, 1961, http://www.jfklibrary.org/AssetViewer/8h9kDI4mik-mYrbo4PJokA.aspx
2. See Berksoy, Özçetin, Wall, this volume.
3. Michel de Montaigne, "On Giving the Lie," in *The Complete Essays*, trans. M. A. Screech (Harmondsworth: Penguin, 1991), 756; Niccolò Machiavelli, *The Prince* (1532), in *The Chief Works and Others, Vol. 1*, trans. Allan Gilbert (Durham: Duke University Press, 1958), 65; Giovanni Botero, *The Reason of State* (1589), trans. P. J. and D. P. Waley (London: Routledge, 1956), 48.
4. See Schrader, this volume.
5. Thomas Hobbes, *Leviathan* (1651), ed. Richard Tuck (Cambridge: Cambridge University Press, 1991), 485.
6. Frank Kitson, *Bunch of Five* (London: Faber and Faber, 1977), 289.
7. Austin Long, *On 'Other War' Lessons from Five Decades of RAND Counterinsurgency Research* (Santa Monica, CA: RAND/Office of the Secretary of Defense, 2006), 53.
8. General Galliéni, *Rapport d'ensemble sur la pacification, l'organisation et la colonisation de Madagascar*, in *The Art of War in World History: From Antiquity to the Nuclear Age*, ed. Gérard Chaliand (Berkeley, CA: University of California Press, 1994), 813-815; Hubert Lyautey, cited in Gwendolyn Wright, *The Politics of Design in French Colonial Urbanism* (Chicago: University of Chicago Press, 1991), 76 and 83; Allard cited in Peter Paret, *French Revolutionary Warfare from Indochina to Algeria: The Analysis of a Political and Military Doctrine* (London: Pall Mall Press, 1964), 30-31.
9. See Schrader, Toews, this volume.
10. *The US Army/Marine Corps Counterinsurgency Field Manual: US Army Field Manual No. 3-24/Marine Corps Warfighting Publication No. 3-33.5* (Chicago: University of Chicago Press, 2006), sects. 5-50 – 5-5-84.
11. See Schrader, Toews, Saberi, this volume.
12. *US Army/Marine Corps Counterinsurgency Field Manual*, sect. 1-153.
13. See, for example, David Galula, *Counterinsurgency Warfare: Theory and Practice* (1964), (Westport, Connecticut: Praeger, 2006), 62; David Kilcullen, "Twenty-Eight Articles: Fundamentals of Company-Level Counterinsurgency" (2006), in Kilcullen, *Counterinsurgency* (Oxford: Oxford University Press, 2010), 43; *US Army/Marine Corps Counterinsurgency Field Manual*, sect. A-45.
14. See Douglas Porch, *Counterinsurgency: Exposing the Myths of the New Way of War* (Cambridge: Cambridge University Press, 2013), 53-54.
15. See Özçetin, this volume.
16. George Orwell, *1984* (1949) (Harmondsworth: Penguin, 1983), 26.

Humpty Putting Himself Back Together
The Hidden Centrality of Participation to Pacification Theory

Stuart Schrader

All the king's horses and all the king's men aren't going to put Humpty together again. Humpty has got to do that himself.

Dennis Duncanson[1]

The critical theory of pacification foregrounds pacification's twinned character. Pacification entails *both* destruction and construction. It yokes development and security. Without pacification's vision of, and tendencies toward, "far-reaching action to construct a new social order," some of the activities associated with pacification would not merit the term.[2] They would be simply destructive, death-dealing, and deleterious. Yet even with the salutary analytic focus on the constructive component of pacification, the linkage between it and pacification's necessarily destructive moments often remains underspecified, undertheorized, or asserted rather than empirically demonstrated. A persistent occlusion occurs when it becomes necessary to identify "who" undertakes the labor of construction beneath the mantle of pacification. Frequently, the passive phraseology of critical analysis illustrates the stumbling block: for example, "schools and clinics were constructed."[3] The destructive agents are readily identified: soldiers, police, intelligence assets, bullets, bombs, missiles, and drones. And, to be sure, soldiers and police often have historically undertaken construction efforts, whether under the yesteryear rubric of "civic action" or the more current one of "humanitarian assistance." But the difficulty for critical analysis of pacification in identifying the hidden "who" of construction is no simple mistake by today's analysts. Instead, this difficulty gets to the heart of how pacification operates. To specify how pacification is constructive is to identify who does the work of construction, on whose behalf, and in what political form. In turn, this specificity enables an understanding of how pacification comprises both destruction and construction.

Rather than treating the question of "who" undertakes construction as an inquiry into syntax, it should be treated as a theoretical question of what we are talking about when we talk about pacification. The pattern of the absent subject originates in the canonical texts of pacification, which were not always clear about who will undertake the labors of developing a new social fabric after the old one has been stretched and torn in the necessary fighting against insurgents, subversives, guerrillas, and so on. For example, Mark Neocleous quotes Joseph Gallieni on Madagascar in 1900:

> As pacification gains ground, the country becomes more civilized, markets are reopened, and trade is re-established. The role of the soldier becomes of secondary importance. The activity of the administrator begins. It is necessary, on the one hand, to study and satisfy the social requirements of the subject people and, on the other hand, to promote the development of colonization, which will utilize the natural resources of the soil and open the outlets for European trade.[4]

The administrator has his work cut out. But the objects of his administrative labors remain unclear. There is a "subject people," but the notion that they have been pacified renders them apparently passive, and thus nearly invisible. They become an object. The present intervention, a fillip to the critical theory of pacification as it now stands, will identify the objects and show them to be active subjects as well. And it will thus answer the "how" question as well.

Without an answer to this who/whom question, the emergent critical theory of pacification struggles to generalize its insights. Indeed, the critical handhold the concept of pacification offers is based on its generalizability, a critical standpoint owing to the generalization, in uneven and tendential ways to be sure, of capitalist social relations that the practice of pacification holds as its ultimate objective. If Neocleous is right that "it is impossible to understand the history of bourgeois society without grasping it as a process of pacification in the name of security and accumulation," pacification holds the possibility for analysts of naming conditions shared across space and time, as well as the mechanisms of their very connection.[5] The pacification concept can exceed the context of counterinsurgencies to cover the operations of the everyday police power deployed in advance of and to stem the need for pointed police intervention.[6] Neocleous cannily points out: "Capital and police dream of pacification: a dream of workers available for work, present and correct, their papers in order, their minds and bodies docile, and a dream of accumulation thereby secure from resistance, rebellion or revolt."[7] But what its analysts dream of is a supple framework for understanding how it is that binary conceptual polarities like war and peace, military and police, fail to accurately capture the conditions of social life under capitalism. To deepen the traction of pacification as a theoretical term, I will wend through empirical details of struggles by its historical advocates to define and delimit it during a period of political upheaval it was intended to address and restrain, namely the US-led pacification programming of the 1960s.

In research on this programming, the answer I have found to the who/whom question is deceptively simple, so simple that perhaps the advocates of pacification sometimes felt little need to be explicit. Yet critical theorists should indulge no such luxury. For it turns out that the people who will do the work of pacification are "the people." To put that another way, the objects of pacification are the subjects of its most important entailment. The constructive work of pacification is a self-help project with a definite spatial locus and a future-oriented temporal status, specifically designed to oblige the target population to invest in its members" own security and controlled uplift. Pacification has historically required the participation of the pacified—that is "how." Otherwise, it is not pacification. It is something else, whether we label it war, destruction, desolation, or genocide. Yet what pacification as a theoretical tool allows is to show that these terms rarely exist unmodified. Their horror rarely stays separate from the hopefulness that accompanies construction and reconstruction. The key is to understand that the objects of such destruction may also be the workers responsible for reconstruction afterward—or, as some pacification experts argued, construction beforehand, as a prelude to clearing the social field of the recalcitrant, malcontents, dissidents, and revolutionaries. Thus, pacification also entailed the consolidation, construction, and often violent homogenization of the "community" to undertake community development, the creation of "the people."

Whether Humpty fell from the wall or was pushed, the experts in pacification agreed that he had to reconstruct himself. Pacification is impossible without participation, and participation takes shape as a most fundamental demand not only from above in liberalism but also from below, in struggles against the shortcomings of, and illiberal carve-outs from, liberalism. The deep embeddedness of the participatory mechanism of pacification within the operating principles of bourgeois order—and within the grammar of its rejection, or at least demands for the fulfillment of its freedom promises—suggests why pacification as a theoretical term can be so fruitful, but it also suggests the perils of its analytic misapplication through a one-sided, undialectical focus on violence alone. Although the critical theory shows that pacification more frequently *does not* than does describe itself using the term, a look at debates among self-conscious advocates of the term and of its implementation, particularly in reference to the US war in Vietnam, elucidates what was at stake in yoking together the two moments—destruction and construction—of pacification in one set of political and economic objectives, while also giving insight into how the debates turned exactly on the who/whom question and the type of action these people would undertake.

PLACING PACIFICATION IN HISTORY AND THEORY

A difficulty endemic to the analysis of pacification derives from its status as a still-live concept and practice. At the same time that the critical theory of pacification is being elaborated, so-called "urban pacification units" are operating on the ground in Brazil, for example.[8] As such, a useful distinction can be drawn between pacification as a category of practice, on the lips of practitioners in the contemporary moment or in the past, and pacification as a category of analysis, applied inductively and conceptually to the contemporary moment or to the past. Another way to put this would be to distinguish between "categories of everyday social experience" and "experience-distant categories used by social analysts."[9] Yet even here a difficulty emerges because analysts of pacification who write for scholarly audiences also have been its practitioners, or they attempt to influence practitioners with their analyses. The critical theory of pacification, however, holds no such ambition of refining pacification. Instead, the critical theory aims to inquire into, disinter, identify, and analytically dissolve the social conditions from which pacification emerges as an imperative and to which its operation gives shape. In this way, the critical theory of pacification is consonant with critical theory writ large, as defined by urban theorist Neil Brenner: "Critical theory is . . . not intended to serve as a formula for any particular course of social change; it is not a strategic map for social change; and it is not a "how to"-style guidebook for social movements." Indeed, invoking Theodor Adorno, Brenner argues that critical theory alights analytically prior to, and abstracts from, Lenin's ever-present question "what is to be done?"[10] Notably, Adorno once quipped that this question, "as an automatic reflex to every critical thought before it is fully expressed, let alone comprehended ... recalls the gesture of someone demanding your papers."[11] The reflex, for him, is itself pacifying, taking its shape and inspiration from the police power. Critical theory thus aims to supersede not only the question of how theory can inform practice but also the social (and theoretical) problematic that gives rise to the question. But that experience-distant moment remains elusive. The clotted history of pacification as such—itself conditioning the social formation within which possibilities of analytic categories are immanent—looms large in analytic attempts to characterize it. As such, because some critical analysts of pacification do aim to offer guidance on how to resist it, an appreciation of the participatory character of pacification by the pacified is crucial. To condemn violence, especially state violence, is the easy part.

To analyze pacification in its historical context, it is important to distinguish bodies of historical literature and draw on them in different ways. First, participants in and advocates of pacification in Vietnam have written their own retrospective analyses of successes and failures.[12] These works overlap with scholarly literature but must be read cautiously, particularly because the overall failure of pacification in Vietnam to achieve a US/Republic of Vietnam victory means that these accounts usually tend to defend the authors" efforts, exculpate them from error, and place blame elsewhere. Yet too much caution in reading these works risks ignoring the insights they offer. Relatedly, the bureaucrats in charge of pacification within the overlapping military, civilian, and intelligence apparatuses (and their overlap is indeed the sine qua non of pacification) produced mounds of records at the time detailing their goals, objectives, tactics, strategies, difficulties, and assessments of outcomes.[13] These mundane nuts-and-bolts documents, I believe, should be integrated more thoroughly into the critical theory of pacification, alongside the more grandiose, programmatic texts that critical scholars have used.

Second, a body of literature on pacification pre-existed 9/11 but was dramatically reshaped and reinvigorated by the subsequent US wars in Afghanistan, Iraq, and beyond of the second Bush and Obama presidencies.[14] The critical theory of pacification itself emerged in this moment and must be seen as itself one of the voices within debates around these wars, though of a notably different tone and perspective.[15] Without the return of pacification to martial vocabularies thanks to renewed military and civilian emphasis on counterinsurgency in Afghanistan and Iraq that retooled older pacification practices, the critical theory arguably would not exist as such. Yet many of the primary proponents and opponents of counterinsurgency, either as active military officers or as professors in military academies, such as Gian Gentile, David Kilcullen, Stanley McChrystal, John Nagl, David Petraeus, and Douglas Porch, have made ample use of historical case studies, historical-comparative analyses, and historical analogy.[16] Although debates concerned how the United States might "win" in Iraq and Afghanistan (or whether the United States should have deployed soldiers to these theaters), the terms of debate often were drawn from a lexicon of historical examples. How one interpreted counterinsurgency in Malaya dictated how one would recommend tactical and strategic decisions for Iraq, for example.[17] As such, critical theory must tread with caution in its use, even repurposing, of the recent pleonastic discussions of counterinsurgency, because the literature so frequently instrumentalizes historical analysis—even if the analysis is occasionally tenable and judicious—for present-day concerns that exceed and submerge the requirements and expectations of historical scholarship. This prescriptive literature conflates a category of practice with a category of analysis.

Third, since 2000 there has been a welter of historical scholarship on "modernization" as well as on its umbrella concept development.[18] Such scholarship treats modernization not as a theory to be applied, as some social scientists continue to do tacitly, but instead treats it as an object of historical investigation whose rise, efflorescence, and collapse can be charted, detailed, and explained. In other words, this historiography has succeeded at distinguishing modernization as a category of practice from modernization as a category of analysis.[19] This work in general focuses on the postwar period (ie, the Cold War) and investigates the intellectual paradigms that accompanied, informed, and enabled the rise of the United States to geopolitical primacy. Recent additions to the literature have included the demonstration of how the modernization paradigm changed in response to conditions on the ground and the excavation of indigenous development schemes that may have been consonant with US-led modernization ideas but did not originate from them.[20] Within this literature sit analyses of pacification, particularly as modernization was a central trope used to understand and organize US action in Vietnam and other areas where insurgency was thought to threaten.[21] Furthermore, recent additions to the literature have been interested in assessing if and how modernization took small-scale forms, in addition to more well-studied massive development projects; how a concern with the small scale and the construction of community as an object of intervention transcended foreign-domestic divides in US policymaking (including pacification programs); and how pacification can be fruitfully applied to the historical analysis of the nexus of security and development projects within and beyond the United States.[22] A persistent division within this historiography has been between research that self-consciously considers itself intellectual history/history of ideas and research that attempts to integrate intellectual history with a focus on the practical implementation of modernization-inflected ideas.[23] The critical theory of pacification risks implicitly recapitulating this divide through its occasionally dehistoricizing deployment of a genealogical method that severs intellectual production from its original conjuncture while stretching historical causality across periods and geographies. And yet at the same time, it is the very ambition of this critical theory to connect the apparently disconnected that is its strength, offering angles of critique on projects of security, development, and pacification that are absent in much of the traditional historiography.

Keeping in mind the productive elasticity of the critical theory of pacification, the term can be adopted more precisely as an analytic category that will still have legs by closer archival attention to its specific genealogy in the context of the US war in Vietnam and the global struggle against Communism. Critical scholarship attempts to highlight the conjoint rise of pacification, colonialism, and liberalism historically across the longue durée.[24] A critical focus on coercion alone, however, instead of on the reciprocity between democratic means and authoritarian means of pacification, risks surrendering its own critique of liberalism's dialectic by dedialecticizing liberalism. This type of critique struggles to apprehend liberalism's attractions, and how formal freedoms engender some of the conditions of possibility for the disorder pacification attempts to corral, while also jettisoning a sense of how it was that Black freedom and anticolonial struggles of the mid-twentieth century, including in Vietnam, pushed liberalism to its breaking point—to a liminal zone of insecurity at which the imperative of security has perennially stood guard. Further, a closer look at how the term pacification was used in the 1960s offers a better understanding of what exactly made the ideas and practices it condenses transportable, so they could be used in the Global South as much as in urban and other areas of the United States.

PARTICIPATION AT HOME AND ABROAD

Insofar as the critical theory of pacification intends to link geographically disparate phenomena, it is important to look for domestic entwinements, overlaps, and parallels with US pacification programming, which is typically understood as having been applied in Third World locales rather than domestically.[25] These are found under the rubric of community development. The US Congress legally enshrined the key mechanism of construction in pacification – participation— but pacification experts were already certain that pacification's only chance for success resided in an active object population. Two pieces of US legislation defined community development in the 1960s at home and abroad: the Economic Opportunity Act (EOA) of 1964, which ushered into being President Lyndon Johnson's Great Society, and the Foreign Assistance Act, as modified in 1966, governing the functioning of the US Agency for International Development (AID). The symmetry between these two pieces of legislation indicates that a commitment to democratic, community-based participation in economic aid spread far beyond the territorial borders of the United States. This commitment shaped a single, multiscalar, global field for development's operation. The transportability of the project resulted from the sense held by development and security experts that the problem of security had no frontiers and that insecurity threatened or precluded development anywhere it arose. Among these experts, those I discuss here are: William E. Colby, of the Central Intelligence Agency (CIA) and Civil Operations and Rural Development Support (CORDS), the centralized directorate for all US counterinsurgency operations in South Vietnam, what it called "pacification," founded in 1967; Charles T. R. Bohannan, a former anthropologist who became a counterinsurgency expert and consulted with the US Army, AID, and various research firms; and Ogden Williams, who also worked with AID in South Vietnam.

To use the word pacification to characterize this transnational development-security project is appropriate, not least because that is the word people such as Colby used. Its applicability derives from the critical theory of pacification's effort to link security and development and the foreign and domestic spheres, a double-linkage that a shared focus on the Economic Opportunity and Foreign Assistance Acts engenders. Demands from below for a say in how community development would proceed at home led to the EOA. These demands also assured a mandate of political participation in foreign aid. Though CORDS was a unique participatory community development operation overseas because it was situated in a country undergoing revolution and counterrevolution, its participation mandate was not unique. US foreign aid at the time enacted participatory development elsewhere, and the mandate should be viewed in the context of a globe wrenched by decolonizing struggles for self-determination.

Domestically, Title II-A of the EOA instituted the goal of "maximum feasible participation" by the recipient community in the management of economic aid. Participation did resonate with some radical visions of democratic self-determination, but in the Johnson administration's formulation it entailed allowing impoverished people a voice in defining what constituted poverty, how it could be managed, and what its mitigation might look like on an everyday level. In contrast to prior modalities of poverty alleviation that prized outside expertise, the program of participation assumed that those who experienced poverty might possess expertise of their own that could address the problem. Expert skepticism of grassroots empowerment could not protect elite-led programs, as grassroots mobilizations had been demanding a say in how poverty alleviation would take shape.

Domestically oriented scholars have faced challenges when attempting elite assent to grassroots mobilization, calling it, for example, "ironic."[26] Here a transnational analysis helps us make better sense of the participation mandate. From the perspective of security professionals, such elite decisions become less surprising and incomprehensible. The homology between participation in the Great Society and in overseas development originated legislatively, but it was already on the minds of security experts. Security concerns were at the forefront in community action's conceptualization and should be seen as integral to its advocates" particular vision of uplift, particularly with regard to juvenile delinquency. Domestic poverty experts, in turn, cast their eyes on unrest in Third World locales to understand what was at stake in controlling insecurity.[27]

A belief in the necessity of some form of participation in counterinsurgency predated the 1966 Foreign Assistance Act, but once in place, CORDS had to report on how well its efforts complied with the participation mandate. Title IX of the 1966 Foreign Assistance Act instituted an almost exact overseas correlate of Title II-A of the EOA. It read:

> In carrying out programs authorized in this chapter, emphasis shall be placed on assuring maximum participation in the task of economic development on the part of the people of the developing countries, through the encouragement of democratic private and local governmental institutions.[28]

This striking parallel under the banner "maximum participation" was paramount in the thinking of those charged with alleviating poverty, and, thus, preventing subversion overseas.

The vision of participation that was to emerge from Title IX consisted of three elements: decision-making, implementation, and reception. The decision-making component included national and local democratic processes, but it also aimed for "participation in decisions that might be considered to be outside the sphere of official public governments," such as, for instance, agricultural cooperatives or trade unions. To achieve decision-making participation, Title IX included "encouragement of democratic institutions and processes; forms of decentralization; and, increased number and effectiveness of voluntary organizations." Second, participatory implementation meant using voluntary (not forced) labor and affording citizens the chance to be "involved in carrying out the decisions they participate in making, with a reasonable hope of obtaining a just share of the benefit." The third element consisted of the rewards of development, "material, cultural, civic, and psychic," that were to be shared among those who participated democratically. Yet, it was warned, "Participation of this kind does not necessarily mean an immediate redistribution of returns among the entire populace." These three pillars evinced liberal optimism while also acknowledging the challenges of inherited political and economic structures in Third World nations.[29]

The limitations US development and security experts encountered overseas were legion, from insufficient language and cultural competencies to labor and equipment shortages. But participation was intended to bypass difficulties while also comporting with a distinctly American liberal ethos of propertied freedom and controlled uplift for those fit for self-government—while enabling the performative demonstration and enactment of this fitness through the very vessel of self-organized participation. What historian Nick Cullather has called "ritualistic displays of allegiance" would demonstrate a tendency toward fitness for self-government even as this loyalty was to be directed to the nation, rather than channeled as political energy into obstreperous demands for self-determination.[30] When the latter occurred, security forces stood primed to respond.

The main practical difficulty to overcome, however, was the blithely dismissed (or anxiously disavowed) perception overseas that the United States was the latest imperial power, building on a palimpsest of prior imperial pursuits, and thus provoking rebellion. The surest way to avoid this problem dovetailed with the practical inadequacies of inexperience and budgetary limitations: to have the people themselves be responsible for development. One dissenting social scientist, Gerard J. Mangone of Syracuse University, alerted an audience of his peers to the potentially rebellion-producing effects of US foreign aid at a March 1962 Army symposium on limited warfare, five years before the inception of CORDS. Mangone contrasted the method-oriented approach of US promotion of promotion of democracy and the guideline of self-help in the Third World with Communism's outcome-oriented appeal. He remarked, "democracy can't tell them what to do or what they should want. They must "discover themselves," whatever that term means." He continued sarcastically, "The United States is not intervening in other people's affairs. All we do is give then billions of dollars to transform their economic system, send abroad missionaries of one kind or another, give them technical assistance that will change their mores, political advice, obtain military bases, use all the propaganda mechanisms at our disposal."[31] The response he received was defensiveness, involving awkward joking and a change of conversation topic. Mangone had put his finger on an uncomfortable reality that did not give much counterinsurgent guidance when counterinsurgents saw themselves as distinctly and unfailingly benevolent. Participation, as the practical texture of the self-help ideal, was the solution.

THE PACIFICATION-PARTICIPATION NEXUS

From the perspective of the advocates of pacification, participatory development became a means of conferring political legitimacy on counterinsurgent state projects—exactly what Mangone saw as lacking. Imposed on them by Congressional legislation, security experts did not ask for Title IX's requirements of participation in development, but they had already recognized the necessity of fostering the population's investment in its own security. For Colby, democratic participation in the management of foreign aid was integral to the goal of political stability. Title IX required that CORDS fashion participatory development plans, but Colby had already been an advocate of them. In fact, his belief in the necessity of participation led to reservations about the term pacification. Colby's skepticism hinged on the implication that a pacified population was an inactive one. He wrote, "We searched around for another name than "pacification," because of its connotation that the population was to be forced into quiescence, when the idea was precisely the opposite, to activate the people in the villages."[32] Participation, for him, would counter "Communist People's War" tactics and extremist appeals by investing the people in their own betterment to assure their loyalty to the government of South Vietnam. Even the CORDS Phoenix program itself fit the mandate, as it enabled targeted "neutralizations" (capture, induced defection, or assassination) of Communists in South Vietnam.[33] Through its encouragement of denunciations and its use of informants Phoenix would enlist the population in its own protection.

Paradoxically, though pacification was officially attached to the civilian side of the war effort, it did not become an acceptable term until after full involvement of the US military, with the arrival of ground troops in South Vietnam and the initiation of aerial bombardment campaigns in the North over the course of 1965. Counterinsurgency impresario Edward G. Lansdale noted that "pacification" was a term avoided by Americans in South Vietnam in the years after 1954 because of its association with the French: " "Pacification" came to have a poor emotional sound to Vietnamese ears, due to some of the security tactics formerly employed."[34] A widely circulated July 1962 list of the US State Department's 18 key terms in "the field of civil defense" did not include "pacification."[35] The word did not come into wide usage by top-level US officials in Vietnam until 1967 or so.[36] Its use before then was largely by officials associated with, supervised, or influenced by Lansdale, based in the rural provinces and responsible for technical assistance. As one social scientist who consulted with AID officials in these provinces wrote in 1966, "the primary responsibility for pacification or counterinsurgency has remained with the Vietnamese. It was intended that the American Provincial Representatives be advisers and teachers, not doers."[37]

Still, military historians have frequently read backward from later phases of the war effort, with the advent of CORDS, when pacification became the preferred term and civilian operations became assigned to the Pentagon's massive budget.[38] This reading gathers nearly everything under the term's umbrella. Richard Hunt writes that pacification "encompassed both military efforts to provide security and programs of economic and social reform and required both the US Army and a number of US civilian agencies to support the South Vietnamese."[39] This definition's invocation of the military is far too imprecise, however. CORDS, as the civilian-overseen locus of pacification, took advantage of the type of civilian, rather than military, counterinsurgent programming the United States had put into place across the Third World in the 1960s. Such programming consisted chiefly of assistance for civil police and community-based economic development efforts, under the aegis of AID and related agencies, including the Peace Corps and the Alliance for Progress.

The possibility of pacification's success also depended on a financial commitment from the US government. Unlike anywhere else, such a commitment was achieved in South Vietnam once the Pentagon's purse opened. With the creation of CORDS beginning in May 1967, which consolidated and reconfigured disparate existing programs, its leaders recognized that pacification as a term felt antiquated, referring to an earlier moment of arm's length, civilian-led or covert support for counterinsurgency in South Vietnam, rather than the militarized turn the US war effort had taken and which CORDS was designed to reshape and retrench.[40] At the same time, the US civilian footprint in South Vietnam had grown decisively alongside the arrival of ground troops and the commencement of aerial bombing of North Vietnam. But the association of CORDS with pacification was apposite in relation to US action in the Third World more generally, with the shooting war directly undertaken by US soldiers in Southeast Asia an anomaly. For CORDS, pacification included programs that AID did, or aimed to, undertake in many other countries: technical assistance geared toward "self-help" and community development in the domains of public health, public works, education, and agriculture.[41] One military historian described the breadth of activities of CORDS as follows in "a listing of the programs and the agencies formerly charged with them":

> New Life Development (AID), Chieu Hoi (AID), Revolutionary Development Cadre (CIA), Montagnard Cadre (CIA), Census Grievance (CIA), Regional and Popular Forces (MACV), Refugees (AID), Field Psychological Operations (Joint US Public Affairs Office), Public Safety (AID), US Forces Civic Action and Civil Affairs (MACV), Revolutionary Develop-

ment Reports and Evaluations (all agencies), and Revolutionary Development Field Inspection (all agencies).

He continued:

> CORDS also assumed coordination responsibility for pacification-related programs of [AID], such as rural electrification, hamlet schools, rural health, village-hamlet administrative training, agricultural affairs, and public works. With few exceptions, all American programs outside of Saigon, excluding American and South Vietnamese regular military forces and clandestine CIA operations, came under the operational control of CORDS.

This historian further notes that General William Westmoreland himself urged that "civil" appear in the appellation of the new command structure that became CORDS.[42]

It is nonetheless instructive to examine how specialists attempted in October 1964 to Americanize the French-sounding term pacification prior to its reappearance in South Vietnam. At the time, Charles Bohannan, who had collaborated with Lansdale in the Philippines, was an independent consultant to AID's offices in South Vietnam, probably through a CIA arrangement. For him, pacification was processual. It consisted of the "processes, actions, and activities necessary, useful, or desirable in rendering an area (which need not be sharply delineated geographically) pacified." For an area to be "pacified" depended on three criteria: "effective government representation," "no significant use of force against the representatives of government," and "no present capability for the organized use of force against the representatives of government." Political control, therefore, was crucial in his view. It was both a means and the end of pacification. But control required the investment of the people in the pacified zone in their own security. He warned that control as a one-sided concept was "counter-productive, for it is at best ludicrous, at worst self-deceiving, to conceive of a government of, by, and for, the people "controlling" the people." Unidirectional state power exercised from above was for him inimical to the American ethos and played into enemy propaganda. Instead, political control could be useful and effective "only when the affected elements help to enforce the controls." Control, in this theory of pacification, required the participation of its beneficiaries.[43] This criterion is the hidden central aspect of pacification theory.

The early rounds of pacification programming in South Vietnam—before the word itself could be safely used—were under greater indigenous control than historians have typically appreciated. These programs also relied on participation. Although various US and British officials have taken credit for the Strategic Hamlets program, for example, historian Edward Miller has recently demonstrated that it issued more directly from Ngo Dinh Diem and his brother, Nhu, building on but departing from their earlier Agroville program. Strategic Hamlets were to be self-sufficient, highly secure communities ruled by a directly elected local council. Especially important to the program's conceptualization were Nhu's readings of the French pacification theorist Roger Trinquier (who invented the term "hameux stratégiques" for such tactics in Algeria).[44] Trinquier, Miller has pointed out, "believed that enlisting the participation of "the people" was essential to success."[45] Yet Nhu, the main thinker behind the Strategic Hamlets program, was also in close contact with Colby. They met weekly for a period.[46]

Although the Strategic Hamlets program never achieved its stated strategic goals, and its failures hastened the Kennedy administration's break with Diem (and Diem's assassination), this moment of failure enabled Colby and others to think more deeply about the necessary parameters of pacification. Colby himself saw pacification as "organizing the countryside to participate in a campaign for security and development, so as to take from the enemy any hope for a base in the population from which to fight the government," and covert CIA efforts during the Kennedy administration that he oversaw, while Diem was still in power, followed from this notion.[47] Moreover, Colby was keenly aware that US support for Diem introduced another layer of hierarchical power that was unlikely to result in compliance among the people. Furthermore, implicit criticisms of the Strategic Hamlets program made by US officials, and explicit ones made by participating Vietnamese people, revolved around its reliance on forced labor. Subsequent efforts frequently faced labor shortages but also aimed at least rhetorically to avoid the pitfall of coercing labor, and Title IX itself would be explicit on AID's need to support labor cooperation. In this vein, Colby wrote, "We had to enlist the active participation of the community in a program to improve its security and welfare on the local level, building cohesion from the bottom up rather than imposing it from top down."[48] The United States could not impose too much top-down power on Diem without angering Diem, but Diem also faced his own delicate problems of legitimacy that bottom-up programming was intended to solve. Although the Diem regime relied heavily on US financial and other material support, it responded to encroachments on its sovereignty with prickly assertions of South Vietnam's unwillingness to become a "rotectorate."[49] The encroachments took the form of demands on how the support be utilized, an increased US troop presence, as well as calls for curtailment of authoritarian practices. From the US perspective, the latter calls, in particular, were intended to help Diem stabilize the legitimacy of his own regime, even as he took the critique as an affront. Participation, as a counter-revolutionary measure to introduce political control from the bottom up, was to cut through these nested hierarchies and the ever-stiffening dissatisfaction of the population, without strengthening the ongoing revolution.

Ogden Williams, another Lansdale associate, put forth an even more explicit view of the relationship of participation to security in a 1964 commentary on a preliminary attempt to define pacification. Williams emphasized conditions in advance of political control. He felt that, as a rough heuristic, to redefine pacification as Build—Secure—Clear was appropriate, which reversed French colonial methods (Clear—Secure—Build) that inevitably faltered after the violent area-clearing phase due to insufficient amounts of personnel for the building phase.[50] Instead, the builders had to be community members, not outsiders. Williams endorsed his agency's commitment to the necessity of "political and psychological action," but he noted that US personnel were constrained in what they could achieve. They could not, for example, act directly to arrange "civil administration, personnel procedures, administration routines" because of their lack of foreign-language faculties and more general experience.[51] Moreover, as Bohannan had recognized, to shape governing processes meant undue US control over the government of South Vietnam. Like the French, the United States was similarly resource-deficient, but the posture of investing the people in the development of their own secure communities would be the prescription.[52] Pacification was premised on the conviction that "expunging the Communist fish from the popular sea," referring to Mao Zedong's most famous aphorism on guerrilla warfare, "must come as a result of a motivated population, not merely an administered one."[53] The final clearing phase, though, was challenging: if hidden enemy forces had been part of the community and participated in the building, then how could they be cleared from the scene and how could the scene be declared secure?

Bohannan, therefore, recognized that the condition of being "pacified" was rarely a steady-state but rather was always in formation and always subject to disintegration. In practice, Build—Secure—Clear and Clear—Secure—Build attempts coexisted. Key, therefore, was Bohannan's definition's implicit use of time and space. Pacification was oriented toward a future status, which was the condition of being pacified, with the recognition that the achievement of that status was likely temporary and constant reapplication of pacification methods would be necessary. Pacification meant prevention. In most of the world there was no major ongoing insurgency, and the United States aimed to prevent the outbreak of insurgency through assistance to civilian police forces, its most widely used modality of counterinsurgency during the Cold War.[54] As proxy instruments to achieve US aims, police were closer to the general population than militaries.

In addition to this temporal orientation, the locus of pacification was a geographical zone, even if, in Bohannan's framing, it was not sharply delineated. In this zone, residents exercised some amount of democratic and participatory control over the key processes of pacification. This geographical zone delimited the community. Bohannan defined the status of "pacified" areas of South Vietnam as lacking a "present" and "organized" capacity for violence addressed to the state and/or newly introducing social and property relations. This criterion, however, was far stricter than what he would have expected at home, as even dormant threats had to be neutralized. A community that policed itself might be able to do so; where it failed, the security forces stepped in. The key difference, therefore, between the crisis situation of South Vietnam and other places, including the domestic United States, resided in counterinsurgency experts" recognition that, in most situations, to stamp out the threat of insurrection entirely was impossible. Civil policing and participatory economic development programming were not intended to do so. Pacification instead was based on cognizance that loyalties and allegiances were "up for grabs," and it could take advantage of all manner of social divisions to reshape allegiances. Pacification assumed, as Bohannan made explicit, resistance. Were there no resistance to liberal property relations and extant political-economic conditions, pacification would be unnecessary. Instead of strict central government control, a relationship of economic transaction oriented toward self-help as well as consensus-based mutual investment was the goal. Participation would ease restoration of order, but it would also allow citizens to air grievances and even mobilize and protest around them. Title IX, like Title II-A of the EOA which it emulated, codified this relationship. Pacification comprised future-oriented, community-based civilian methods of the prevention of violence and maintained that ongoing prevention as its goal. The most mobile aspects of pacification thus were not the most abjectly coercive, but rather those that attempted to further participation and police-enforced rule of law.

CONCLUSION

Title IX of the Foreign Assistance Act goes largely unmentioned in the historiography of development, modernization, or pacification. Yet it was a key workshop for the elaboration of a resolutely transnational vision of the mechanics of development on an American model. It not only paralleled the community-participation mandate of the EOA, but it also created the possibility of, if not the demand for, concrete transfers across borders. An analytic focus on Title IX, in fact, undermines prevailing scholarly understandings of territorially bounded poverty knowledge, or its outward diffusion from the United States to its peripheries. Perhaps no statement better illustrates the hopefulness and expansive vision Title IX entailed than this line from a 1968 conference on its implementation: "The fact that the US has been unable to resolve its own city problems should not lead Americans to stay out of this area in the Third World."[55] Pacification, therefore, was not a refined practice, perfected domestically for export overseas. Nor was the reverse true. Experts in counterinsurgency were well versed in the necessity of stitching together coercive and participatory projects if the aim were development even before Johnson inaugurated the Great Society. But experiments in Vietnam and elsewhere could not be repatriated unaltered. Instead, solutions would be found through the articulation of these spheres, in the recognition of legal differences and practical similarities between them. It may therefore be argued that what the critical theory of pacification demands in terms of expansive and ambitious research questions, Title IX quietly enforced.

Pacification as a critical concept, as a category of analysis, can be better applied to a wide range of objects of analysis if the specificities and difficulties of its development as a category of practice in the 1960s are taken into account. The key aspects of pacification then, which found their domestic echoes in community-action legislation that many in the neoliberal moment now recall with a wistful eye, were its future-oriented, community-centered processes of participatory investment of the object population in its own security for the purposes of achieving legitimacy for the developmental state to ward off violent challenges to state authority. Attention to these mobile and transnational aspects of pacification will enable a better application of the term today to a variety of situations. The prescriptive literature of pacification has erased the role of participants in their own self-help. This erasure should be interpreted as a disavowal not of their importance to the project of pacification but of the tendency of pacification to fail. The failure results from self-help self-transforming into self-determination. The problem for opponents of pacification is that its failure so frequently demands only further rounds of its application.

ACKNOWLEDGMENTS

I am grateful to the participants in the Law-Capital-Pacification stream at the Critical Legal Conference at the University of Sussex in September 2014, who responded to the first presentation of some of these ideas and from whom I have received generous feedback. Special thanks to Tyler Wall and Mark Neocleous for their guidance.

NOTES

1 Duncanson was a counterinsurgency expert assigned to the British Advisory Mission in Vietnam and then to the British Embassy in Saigon. This quote comes from a conversation with journalist Richard Critchfield in spring 1966; Richard Critchfield, *The Long Charade: Political Subversion in the Vietnam War* (New York: Harcourt, Brace & World, 1968), 283.
2 Mark Neocleous, George Rigakos, and Tyler Wall, "On Pacification: Introduction to the Special Issue," *Socialist Studies/Études Socialistes* 9, no. 2 (2013): 1-6, 1.
3 Neocleous, Rigakos, and Wall, "On Pacification," 7-8.
4 Mark Neocleous, "Security as Pacification" in *Anti-Security*, eds. Mark Neocleous and George S. Rigakos (Ottawa: Red Quill Books, 2011), 23-56, 29.
5 Mark Neocleous, "The Dream of Pacification: Accumulation, Class War, and the Hunt," *Socialist Studies/Études Socialistes* 9, no. 2 (2013): 7-31, 9.
6 The critical theory of pacification grows out of what one collection calls "the new police science," and it shares with it many methodological and theoretical moves, as well as authors, including Neocleous. See Markus D. Dubber and Mariana Valverde, eds., *The New Police Science* (Stanford, CA: Stanford University Press, 2006). This pacification literature is particularly successful at joining two dominant epistemological strands in critical legal/criminological research: the Foucauldian or genealogical and the Marxist.
7 Neocleous, "The Dream of Pacification," 18.

8 Janet Tappin Coelho, "Brazil's "Peace Police" Turn Five: Are Rio's Favelas Safer?" *Christian Science Monitor*, December 19, 2013; Sebastian Saborio, "The Pacification of the Favelas: Mega Events, Global Competitiveness and the Neutralisation of Marginality," *Socialist Studies/Études Socialistes* 9, no. 2 (2013): 130-145.
9 Rogers Brubaker and Frederick Cooper, "Beyond "Identity"," *Theory and Society* 29, no. 1 (2000): 1-47, 4.
10 Neil Brenner, "What is Critical Urban Theory?" *City* 13, no. 2 (2009): 198-207, 201-2.
11 Theodor W. Adorno, "Marginalia to Theory and Praxis" in *Critical Models: Interventions and Catchwords*, trans, Theodor W. Adorno and Henry W. Pickford (New York: Columbia University Press, 2005), 259-278, 276.
12 For example, William E. Colby and Peter Forbath, *Honorable Men: My Life in the CIA* (New York: Simon & Schuster, 1978); Robert W. Komer, *Bureaucracy Does Its Thing: Institutional Constraints on US-GVN Performance in Vietnam* (Santa Monica, CA: RAND Corporation, 1972); http://www.rand.org/pubs/reports/R967; Robert W. Komer, "Impact of Pacification on Insurgency in South Vietnam," *Journal of International Affairs*, 25, no. 1 (1971): 48-69.
13 A place to begin is Edwin E. Moïse's extensive online bibliography of primary and secondary sources. The pacification section is located here: http://www.clemson.edu/caah/history/facultypages/EdMoise/villages.html.
14 To adequately list works on Vietnam would be impossible. For a major pre-9/11 work on pacification, see Richard A. Hunt, *Pacification: The American Struggle for Vietnam's Hearts and Minds* (Boulder, CO: Westview Press, 1995).
15 The two collections that define the critical theory of pacification are: Mark Neocleous and George S. Rigakos, eds., *Anti-Security* (Ottawa: Red Quill Books, 2011), and the "On Pacification" issue of *Socialist Studies/Études Socialistes* 9, no. 2 (2013). See also Mark Neocleous, *The Fabrication of Social Order: A Critical Theory of Police Power* (London: Pluto Press, 2000).
16 See Thomas E. Ricks, "The COINdinistas" *Foreign Policy*, November 30, 2009, http://www.foreignpolicy.com/articles/2009/11/30/the_coindinistas. Nagl and Gentile define the modal proponent and opponent positions, respectively: John Nagl, *Learning to Eat Soup with a Knife: Counterinsurgency Lessons from Malaya and Vietnam* (Chicago: University of Chicago Press, 2005); Gian Gentile, *Wrong Turn: America's Deadly Embrace of Counterinsurgency* (New York: The New Press, 2013). See also David Hunt, "Dirty Wars: Counterinsurgency in Vietnam and Today," *Politics & Society* 38, no. 1 (2010): 35-66; Toews, this volume.
17 A savvy rejoinder to this literature is Hannah Gurman, ed., *Hearts and Minds: A People's History of Counterinsurgency* (New York: The New Press, 2013).
18 A useful brief historiographical overview is: Daniel Immerwahr, "Modernization and Development in US Foreign Relations," *Passport: The Society for Historians of American Foreign Relations Review* 43, no. 2 (2012): 22-25, https://shafr.org/sites/default/files/Passport-September-2012.pdf. Signal works are Latham, *Modernization as Ideology: American Social Science and "Nation Building" in the Kennedy Era* (Chapel Hill: University of North Carolina Press, 2001), which includes a chapter on the Strategic Hamlets program in South Vietnam; Nils Gilman, *Mandarins of the Future: Modernization Theory in Cold War America* (Baltimore: Johns Hopkins University Press, 2003).
19 Nick Cullather demonstrates the value of this approach in "Development? It's History," *Diplomatic History* 24, no. 4 (2000): 641-653.
20 Bradley R. Simpson, *Economists with Guns: Authoritarian Development and US-Indonesian Relations, 1960-1968* (Stanford: Stanford University Press, 2008); Edward G. Miller, *Misalliance: Ngo Dinh Diem, the United States, and the Fate of South Vietnam* (Cambridge, MA: Harvard University Press, 2013).
21 See Joy Rohde, *Armed With Expertise: The Militarization of American Social Research during the Cold War* (Ithaca, NY: Cornell University Press, 2013).
22 Miller, *Misalliance*; Daniel Immerwahr, *Thinking Small: The United States and the Lure of Community Development* (Cambridge, MA: Harvard University Press, 2015); Alyosha Goldstein, *Poverty in Common: The Politics of Community Action during the American Century* (Durham: Duke University Press, 2012); Sheyda Jahanbani, " "Across the Ocean, Across the Tracks": Imagining Global Poverty in Cold War America," *Journal of American Studies* 48, no. 4 (2014): 937-974; Jahanbani, "One Global War on Poverty: Fighting "Underdevelopment" at Home and Abroad, 1964-1968," in *Beyond the Cold War: The United States and the Global Challenges of the 1960s*, eds. Mark Lawrence and Francis Gavin (New York: Oxford University Press, 2014), 97-181; Amy C. Offner, "Anti-poverty Programs, Social

	Conflict, and Economic Thought in Colombia and the United States, 1948-1980" (PhD diss.: Columbia University, 2012); Ananya Roy, Stuart Schrader, and Emma Shaw Crane, " "The Anti-Poverty Hoax": Development, Pacification, and the Making of Community in the Global 1960s" *Cities* 44 (2014): 139-145; Roy, Schrader, and Shaw Crane, "Gray Areas: The War on Poverty at Home and Abroad," in *Territories of Poverty: Rethinking North and South,* eds. Ananya Roy and Emma Shaw Crane (Athens: University of Georgia Press, 2015), 289-314.
23	See Nils Gilman, "H-Diplo Article Review of "Special Forum: Modernization as a Global Project," *Diplomatic History* 23, no. 3 (June 2009), http://www.h-net.org/~diplo/reviews/PDF/AR238-A.pdf.
24	See Neocleous, "Security as Pacification"; Neocleous, "The Dream of Pacification"; George S. Rigakos, " "To Extend the Scope of Productive Labor": Pacification as a Police Project" in *Anti-Security,* eds. Neocleous and Rigakos (Ottawa: Red Quill Books, 2011), 57-83. See also Saberi, this volume.
25	For the injunction to use the critical theory of pacification to link the foreign and the domestic, as well as an illustration of how genealogical method might draw connections across centuries, see, Rigakos, " "To Extend the Scope of Productive Labor"," 62-3, 66-78. Some of the arguments and evidence in this section and the next appear in different form in Stuart Schrader, "To Secure the Global Great Society: Participation in Pacification" *Humanity: An International Journal of Human Rights, Humanitarianism, and Development* 7, no. 2 (2016): 225-253.
26	Noel A. Cazenave, *Impossible Democracy: The Unlikely Success of the War on Poverty Community Action Programs* (Albany: State University of New York Press, 2007), 13.
27	On linkages between counterinsurgency and juvenile-delinquency experts, see, Roy, Schrader, and Shaw Crane, "Gray Areas" and " "The Anti-Poverty Hoax"."
28	Public Law 89-583, September 19, 1966; PL 89-583; Reports on Enrolled Legislation, 9/12/66-9/20/66, Box 40; Lyndon Baines Johnson Library, Austin, TX. For the legislative history of Title IX as well as subsequent amendments, see, Brian E. Butler, "Title IX of the Foreign Assistance Act: Foreign Aid and Political Development," *Law & Society Review* 3, no. 1 (1968): 115-152.
29	Max F. Millikan, Lucian W. Pye, and David Hapgood, *The Role of Popular Participation in Development: Report of a Conference on the Implementation of Title IX of the Foreign Assistance Act, June 24 to August 2, 1968* (Cambridge, MA: MIT Press, 1969), 23-27.
30	Cullather's observation comes in a discussion of the Strategic Hamlets program, investigated further below, which relied on, as displays, "the daily reporting on the radio, the mustering of guards, the mending of barbed wire, and the turning of a wooden arrow, atop a watchtower, to steer helicopter pilots to the location of Viet Cong attackers." Nick Cullather, " "The Target Is the People": Representations of the Village in Modernization and US National Security Doctrine," *Cultural Politics* 2, no. 1 (2006): 29-48, 41.
31	William A. Lybrand, ed., *Proceedings of the Symposium "The US Army's Limited-war Mission and Social Science Research," March 26-28, 1962* (Washington, DC: Special Operations Research Office, American University, 1962), 238-239, with responses on 244-247.
32	Colby and Forbath, *Honorable Men,* 256.
33	John Prados, *Lost Crusader: The Secret Wars of CIA Director William Colby* (New York: Oxford University Press, 2003).
34	Memo, Informal Discussion with Col. Lansdale, USAF, and Col. Valeriano, Philippine Military Attache, 22 February 1957, February 27, 1957; Attachment, Comments on Memo of 27 Feb 57; Military Policy in Underdeveloped Areas; Papers of Edward G. Lansdale, Box 80; Hoover Institution Archive, Stanford, CA (HIA).
35	Airgram CA-478, Dean Rusk to All Diplomatic Posts, July 12, 1962; Special Group (CI) 7/62-11/63; Meetings and Memoranda, National Security File, Box 319; John F. Kennedy Library, Boston, MA.
36	Richard A. Hunt gives 1964–1965 as the time when "pacification" became a more widely used term, citing the CIA officer Douglas Blaufarb. Nhu had already begun using it, convening a Pacification Committee, around 1962. But the point is that its deployment by high-level US officials relied on its previous usage by South Vietnamese officials and those CIA and AID officials charged with assisting them in motivating and directing Vietnamese-led efforts. Hunt, *Pacification,* 3.
37	George K. Tanham, "Challenge and Response," in Tanham, ed., *War Without Guns: American Civilians in Rural Vietnam* (New York: Frederick A. Praeger, 1966), 29.

38 The "R" in CORDS originally stood for "Revolutionary," but it was changed to "Rural" in 1970. This 1970 shift signaled that ideas of Robert W. Komer, the fervid first head of CORDS who was responsible for a prior shift from "rural development" to "revolutionary development" cadres as the appellation of "embedded development agents" dispersed throughout the Vietnamese countryside, had fallen out of favor. It also signaled waning assurance that modernization, entailing a socioeconomic "revolution" harnessed to US-friendly, liberal and democratic ends, was the objective. On CORDS and the "crisis of modernization," see Christopher T. Fisher, "The Illusion of Progress: CORDS and the Crisis of Modernization in South Vietnam, 1965-1968," *Pacific Historical Review* 75, no. 1 (2006): 25-51; The phrase "embedded development agents" is on 36.
39 Hunt, *Pacification*, 2. A lucid accounting of civilian-led pacification programming in South Vietnam completed just as the shift toward the creation of CORDS (and thus the widespread use of the term, particularly by military forces) was set into motion is William A. Nighswonger, *Rural Pacification in Vietnam* (PhD diss., American University, 1966). Nighswonger also contributed to Tanham, *War Without Guns*.
40 See, Fisher, "The Illusion of Progress," 42.
41 Pacification (A Reference Pamphlet), March 1969; Folder 7; William Colby Collection, Box 3; The Vietnam Center and Archive, Texas Tech University, Lubbock, TX; http://www.vietnam.ttu.edu/virtualarchive/items.php?item=0440307001.
42 Thomas W. Scoville, *Reorganizing for Pacification Support* (Washington DC: Center of Military History, US Army, 1982), 67, 62. MACV means Military Assistance Command—Vietnam.
43 Memo, Charles Bohannan, Counter-Insurgency Terms, Objectives and Operations, October 23, 1964; Photocopies of Memoranda and Reports on Vietnam, 1963-1964; Charles T. R. Bohannan Papers 1915-1985, Box 1; HIA.
44 Miller, *Misalliance*, 231-9.
45 Miller, *Misalliance*, 233.
46 Prados, *Lost Crusader*, 73.
47 Colby and Forbath, *Honorable Men*, 256.
48 Colby and Forbath, *Honorable Men*, 176.
49 Miller, *Misalliance*, 229.
50 See Toews, this volume.
51 Comments by Ogden Williams on ToAid A-822 "Definition of Pacification" dated October 8, 1964, October 28, 1964; ArBoso; Charles T. R. Bohannan Papers 1915-1985, Box 2; HIA.
52 See Saberi, this volume.
53 Memo, Richard Helms to Robert W. Komer, Report to the President on Your Recent Trip to Vietnam, July 1, 1966; CIA Records Search Tool, Document No. CIA-RDP80B01676R000100060007-0.
54 Stuart Schrader, "American Streets, Foreign Territory: How Counterinsurgent Police Waged War on Crime" (PhD diss., New York University, 2015).
55 Millikan, Pye, and Hapgood, *The Role of Popular Participation*, 10.

Counterinsurgency as Global Social Warfare[1]

Ryan Toews

> *... an investigation of capitalistic militarism would bring to light the most deeply hidden and delicate root-fibres of capitalism. Again the history of militarism would be the history of the strained relations and jealousies between nations and states, arising from their desire for political and social power or economic advantage; at the same time it would be the history of class-struggles within nations and states for the same objects.*
>
> Karl Liebknecht[2]

Over the past decade, the American military's experience of occupation in Afghanistan and Iraq renewed emphasis on counterinsurgency warfare. Key to this new emphasis is a coterie of military intellectuals who sought to reorganize military doctrine and capacity for counterinsurgency, not just in Afghanistan and Iraq, but for the so called 'Long War,' a war which extends temporally and geographically well beyond Iraq and Afghanistan to include the largely asymmetric conflicts with other (and future) 'rogue' states and insurgent forces, often framed in civilizational terms.[3] As one counterinsurgency proponent explained, "To meet future challenges [posed by small wars], America's Army must turn from the warm and well deserved glow of its Persian Gulf victory and embrace, once more, the real business of regulars, the stinking gray shadow world of 'savage wars of peace' as Rudyard Kipling called them."[4] Much of this work on doctrine was developed in a revised American counterinsurgency manual, released in 2006, as well as broader military and academic writing on counterinsurgency. Over the subsequent years, this approach would aim to significantly reshape the American military's thinking about the wars it should be prepared to fight and how it would fight them, breaking from the military's emphasis on the Weinberger doctrine of 'decisive victory' and the technology centred 'Revolution in Military Affairs' that would inform the Bush administration's invasions of Afghanistan and Iraq. This intellectual and doctrinal shift is the focus of this chapter.

While more critical scholarship recognizes this shift to counterinsurgency as part of a broader project to maintain and extend US imperialism, the perspective of counterinsurgency's proponents would also often be an unapologetic defense of US imperialism. Counterinsurgency may be framed as a method of stability operations, peace building, or as a reluctant US leadership coping with a fragmenting world.[5] Nonetheless, a sense of an imperial role, and an imperial lineage that connects the US to the earlier imperialisms of Britain and France, is usually present in the counterinsurgency literature, even if it is supposedly a kinder, gentler imperialism, or an Imperialism Lite.[6] For Sarah Sewell, a contributor to the 2006 Counterinsurgency Manual and at the time the director for the Carr centre for Human Rights at Harvard, the new US doctrine 'may draw upon colonial teachings and the US Marine's code of conduct for occupying Latin American nations. But its implicit and explicit standards of behavior have evolved. The new manual is cognizant of international rights standards, expectations of accountability, and the transparency that accompanies the modern world.'[7] Montgomery McFate, another manual author, would explicitly call her own discipline of anthropology to return to its historic role as a "handmaiden of colonialism"[8] and its primary task of "translating knowledge gained in the 'field' back to the West,"[9] through working for US counterinsurgency efforts. Max Boot, a proponent of the new counterinsurgency, titled a 2002 book celebrating the US role in colonial wars and occupation *The Savage Wars of Peace*. This idea and Boot's title are noted approvingly by Thomas Nagl, another one of the COIN manual authors whose own book *Learning to Eat Soup with a Knife* explained what the Americans in Vietnam should have learned from British colonial warfare in Malaya.[10]

Such comments are fairly indicative of the writings of other American counterinsurgency proponents who drew almost exclusively from the efforts of the British and the French to extend or maintain colonial rule, as well as US experiences in Central America, the Philippines, the Caribbean, and Vietnam. The work of 1960s era French counterinsurgency theorist David Galula is resurrected for the 'Long War.' Joining Galula are other Western colonial officers and 'experts' with experience in the colonies including Lawrence of Arabia and Robert Thompson of the British colonial experience, and Hubert Lyautey who helped oversee French colonial expansion in Africa.[11]

In this chapter I tackle these imperial assumptions in the 'new' counterinsurgency doctrine. In focusing on the political economy of American imperialism, much of the Marxist literature has tended to treat the military as simply an extension of state power rather than understand military doctrine and knowledge production as an important object of analysis. This is a mistake, and it reflects a weakness in the theorizing of imperialism. Consequently this chapter argues that the American military's development of a counterinsurgency doctrine provides important insights into the contemporary nature of American imperialism. It draws from Mark Neocleous's critical reading of the concept of 'pacification' to link contemporary American counterinsurgency doctrine to the internationalization of capital.

The question of counterinsurgency's role within imperialism is a bigger one than can be adequately addressed here. Consequently the scope is narrowed to specifically the counterinsurgency doctrine and the intellectual work around it, and not to its specific application in Iraq, Afghanistan, or elsewhere. The chapter is organized into four sections. First, I discuss Marxist accounts of imperialism, specifically focusing Neocleous's critical reading of pacification. Second, I explain the emergence of a 'new' counterinsurgency doctrine within the US military over the past decade. Third, I discuss the characteristics of the 'new' counterinsurgency with an emphasis on both the linkages to colonial warfare and also how it is captured by a critical reading of pacification. Fourth, I conclude by suggesting that we need to think of counterinsurgency as 'social war' in a Marxian sense, a war to secure the conditions of accumulation on a global scale.

IMPERIALISM, ACCUMULATION, PACIFICATION

The most immediate historical reference point for modern imperialism is the direct occupation over vast sections of the globe by European nation-states, an occupation that intensified in the late nineteenth century, and which declined in the period following World War Two. The kernel of a Marxist analysis of imperialism was already present in Marx's *Communist Manifesto* in 1848:

> The need of a constantly expanding market for its products chases the bourgeoisie over the whole surface of the globe. It must nestle everywhere, settle everywhere, establish connexions everywhere. ... It compels all nations, on pain of extinction, to adopt the bourgeois mode of production; it compels them to introduce what it calls civilisation into their midst, i.e., to become bourgeois themselves. In one word, it creates a world after its own image.[12]

Much in Marx's analysis would be amended by subsequent thinkers from Hilferding and Lenin onwards – but Marx's analysis contained the essence of a critique of imperialism. The rapacious expansion of the European and American colonial empires beginning in the late nineteenth century, two world wars, and the subsequent anti-colonial challenge made for a rich and evolving theorization of imperialism through the twentieth century.[13]

Opposed to the apologies for imperialism that presented it as a 'civilizing' (or, later, 'democratizing') mission, Marxists emphasized that imperialism was driven by the logic of capital.[14] The compulsion to accumulate 'chases the bourgeoisie' in Marx's words "over the whole surface of the globe." The consequence was "a world after its own [the bourgeoisie's] image," a world that was profoundly geographically unequal and politically dominated by Western Europe, and later Japan and the United States.

In the aftermath of European colonialism, the term 'globalization' was used to describe the growing, albeit unequal, interpenetration of national economies, and the growing importance of finance capital, and led to a rethinking of imperialism. Rather than being operationalized through direct occupation, imperialism is conceived of as operating through the institutions of the global economy such as the IMF, the World Bank, the WTO, multi-national corporations, and through states via free trade agreements, the imposition of neo-liberal restructuring. But an account of imperialism that locates it solely in the processes of globalization could not explain the shift in the 2000s towards direct US occupation in Iraq and Afghanistan, or drone strikes in Pakistan or Yemen, or the global spread of US military bases.

In the period following the American led invasions of Afghanistan and Iraq, there were a number of attempts to theorize this 'new' imperialism that could link American and, more broadly, Western military violence to a new strategy of accumulation.[15] One of the most influential was David Harvey's conception of 'accumulation by dispossession.'[16] Harvey drew this concept from a distinction between two forms of accumulation in Marx's writing, accumulation by expanded reproduction and what Marx called 'primitive' (or 'original') accumulation. The former encapsulates the standard Marxist account of capitalist accumulation. This appearance of a free exchange between individuals in the market masks an underlying expropriation of surplus value based on the real inequality between the owner of money and the owner of labour-power. For Marx, the original source of the money-power of the capitalist lay in an 'original' accumulation, an historical process that combines the appropriation of wealth through the early period of colonialism with the dispossession and proletarianization of the peasantry in the Western European and specifically English countryside. This historical process creates for Marx the two classes of capitalism, the bourgeoisie, and the proletariat. For subsequent critics of imperialism, it also produced the geographic division of the world into a centre and periphery.

Harvey suggests that original accumulation was, for Marx, a form of 'original sin,' a pre-history of capitalism. Harvey, however, argued that this accumulation by dispossession is a continuing process, intrinsic to capitalism, and driven by the over-accumulation of capital—capital that is continually searching for new sources of profit. It entails the commodification of previously un-commodified or de-commodified features of human society or the capture of ownership over particular commodities. These new commodities did not enter the market through 'free' exchange but were appropriated through directly political means.[17] Harvey placed American military power at the centre of a theory of imperialism as a spatial strategy of accumulation, but as with many other Marxist critics of imperialism, this did not lead to an analysis of military doctrine or practices. Although it is suggested that the military would have a role in constructing the social conditions for accumulation, the issue is not systematically explored. Of course, Harvey's account of accumulation of dispossession precedes the official re-emergence of counterinsurgency doctrine within the US military, and he is concerned with the role of the military as an instrument of US policy in creating the conditions for capital accumulation through increasing American dominance in the Middle East and to contain the resistance to the displacements caused by accumulation, what Klassen calls a disciplinary militarism.[18] But it leaves unexplored the ways that military practices themselves are conducive to accumulation.

The critical reading of pacification being developed in this volume and elsewhere provides a way to address this. For Neocleous, pacification is integral to accumulation and for reasons that are more than simply the crushing of resistance.[19] Neocleous emphasizes the way that Marx's account of 'primitive accumulation' contained a recognition of a productive use of violence in creating institutions that enable accumulation and the right kind of workers for it, not simply as a prior history but as a continuous process.[20] Building from Marx, Neocleous uses 'pacification' to capture this dynamic of "a war to build rather than destroy." He writes that the "key practice of pacification is nothing less than a feat of enormous social engineering to (re)build a social order. And what is to be built in this new order is a secure foundation for accumulation."[21] What this is characterized by is "a dream of workers available for work, present and correct, their papers in order, their minds and bodies docile, and a dream of accumulation thereby secure from resistance, rebellion or revolt."[22] Neocleous's use of pacification addresses more broadly the necessity of state violence (and he rejects the conceptual separation between the police power and the war power) to capital's desire to accumulate.[23] In doing so, he challenges the conception of liberal peace noting that the periods of 'peace' have been bound up in processes of force linked to creating the conditions of capital accumulation.[24] This reading of pacification captures a key feature of contemporary counterinsurgency warfare and its precursors in colonial warfare, in that it is a social war whose end is the construction of a new social order, one in which imperialism's aim of accumulation is secured.

Other critical literature has examined the massive appropriation of wealth that has followed from the occupations in Iraq and Afghanistan. As is well known, the enormous defense expenditures required have been a mechanism to transfer public wealth into private hands through companies such as Haliburton, Bechtel, Lockheed and others through the provision of weapons, infrastructure, and personnel for the occupations.[25] The privatization of public assets has also followed from occupation. Naomi Klein has documented how Iraq's occupation was an enormous experiment in neoliberal restructuring and privatization, which decimated, among other things, social welfare provision, labour, and the public sector.[26] Resource wealth was transferred in both Iraq and Afghanistan to American and other multinationals, and this has often been identified as the leading cause of the wars.²⁷ Central Asian economies have also been opened up to global capital with Afghanistan becoming a piece of a new 'Silk Road.'²⁸

As important as these moments of accumulation are to American imperialism, it is not what this chapter directly focuses on. In comparison to the requirements of conventional war, counterinsurgency is war on the cheap, and the new counterinsurgency largely came into being after the sell-off of Iraq and Afghanistan. Rather the problem is the construction of a legitimate state and society, conducive and open to capital. Counterinsurgency is about capital's dream for a globally integrated market, docile populations, and local institutions that can enforce bourgeois law for the purposes of accumulation. In his own defense of American empire, Niall Ferguson argued that "Capitalism and democracy are not naturally occurring, but require strong institutional foundations of law and order. The proper role of an imperial America is to establish these Institutions where they are lacking, if necessary … by military force."²⁹ Ferguson's take on American Empire is neither unique nor inconsistent with official American national security objectives that identify the global economy and market access as core national security concerns.³⁰ Consequently, Klassen, among others, suggests that we should read American imperialism as a project of hegemonic liberalism aimed at expanding "capitalism on a global scale and [embedding American] primacy in worldwide structures of economic, political, and military power."³¹ For this project, counterinsurgency is more than simply a disciplinary militarism, but one that intends itself to be productive of particular institutions and social relations that will integrate 'the barbarians' into orbit of global capital.³² I shall develop this point thorough an account of the recent return of counterinsurgency.

THE 'RETURN' OF COUNTERINSURGENCY

In December of 2006, the US military published a new counterinsurgency manual that was then republished to great fanfare in July of 2007 by the University of Chicago Press. It was updated again in 2014 but it is worth discussing the origins of the 2006/2007 edition. The manual was written by a group of US officers and American academics overseen by David Petraeus, at the time a rising star in the US military and the architect of the surge that was perceived to have turned the tide of the US occupation in Iraq. Other participants included lead author Dr. Conrad Crane, an American military historian, Australian counterinsurgency expert David Kilcullen, the aforementioned Thomas Nagl, Montgomery McFate, and Sarah Sewell, and many more linked to both the military and academia.³³ In keeping with the claim that counterinsurgency was the graduate level of warfare, one of the things that distinguished the writers was their social science training.

The authors' motivations were to update counterinsurgency doctrine for the American military and overcome its institutional resistance to counterinsurgency. From this perspective, the US had emerged out of the Cold War prepared to fight a modern war in Europe against the USSR. Although it had engaged in counterinsurgency (and insurgency) warfare elsewhere, it had failed badly at it in Vietnam. For the authors, the defeat in Vietnam was a consequence of the failure to apply effectively the right counterinsurgency approach, a failure brought about by the military's obsession with a conventional war in Europe and their unwillingness to treat Vietnam differently.[34] The legacy of this was to turn away from counterinsurgency warfare until well after the occupation in Iraq, leading, for counterinsurgency proponents, to 'failure' in Iraq and Afghanistan. Jack Keane, former Vice Chief of Staff of the Army explains this narrative, "[in Iraq] we put an army on the battlefield that ... doesn't have any doctrine, nor was it educated and trained, to deal with an insurgency ... After the Vietnam War, we purged ourselves of everything that had to do with irregular warfare or insurgency, because it had to do with how we lost that war. In hindsight, that was a bad decision."[35] From this narrative, the US military was unwilling and unprepared to fight a counterinsurgency war in Iraq and Afghanistan.

This rejection of counterinsurgency is exaggerated given American counterinsurgency efforts in Central America, Afghanistan and elsewhere in the Cold War (and El Salvador is frequently cited as a success story by these same counterinsurgency proponents),[36] but it is crucial to the counterinsurgent narrative of a US military that has to be transformed for the wars of the future. David Ucko, among others, notes that well into the wars in Iraq and Afghanistan, the US military did not recognize the type of war it was fighting, a view repeated by other counterinsurgency proponents.[37] In Ucko's account, the US commitment to the Weinberger doctrine and the 'Revolution in Military Affairs' in the 1990s with their emphasis on conventional war and technological solutions squeezed out counterinsurgency and prevented an understanding of the wars that the US would come to fight.[38]

Only after the failures on the ground in Iraq made the necessity of counterinsurgency apparent and the right leadership, in David Petraeus and his coterie of highly educated military and civilian advisors, emerged to apply the new doctrine, was the Iraq occupation 'saved.' In doing so, Petraeus and his supporters overcame the institutional commitment to a conventional approach to warfare not appropriate to the types of warfare that the US would have to fight in the 'Long War.' From this standpoint, success in Iraq, Afghanistan, and other potential sites of the 'Long War' depends both on a return to counterinsurgency and on getting it right. Gian Gentile, a conservative critic of the 'new' counterinsurgency, noted six years after the publication of the counterinsurgency manual that despite having little basis in the historical record, this narrative permeates the intellectual culture of the US military.[39] As a consequence, in the intervening period the US military significantly expanded its counterinsurgency capacities.

US proponents of the approach identified in the manual usually highlight David Galula as the key theorist informing the 'new' population centric approach. Max Boot labeled him the 'Clausewitz of counterinsurgency'[40] and someone who transformed contemporary US thinking about counterinsurgency, based on two books he wrote in the 1960s while working at Harvard, a job arranged by Vietnam War Commander Westmoreland.[41] Galula had obtained significant experience of counterinsurgency by this time. He fought with the free French forces in World War Two, and was subsequently stationed to Beijing with a French Military attaché where he immersed himself in the Chinese Civil War, and he was briefly captured by the Chinese Red Army. Following this, he was a military observer during the Greek Civil War, where he viewed the 'successful' application of counterinsurgency by the American- and British-backed Greek military. He closely observed the French war in Vietnam while stationed in Hong Kong, and he applied his experience in the French pacification efforts in Algeria in the late 1950s. For his advocates, these experiences made him uniquely able to understand counterinsurgency.[42] While Galula subsequently faded into obscurity after Vietnam, his counterinsurgency writing was approvingly noted in Andrew Krepenivich's account of the US failure to apply counterinsurgency in Vietnam, *The Army in Vietnam*. The text was influential for the authors who updated the manual, and it shaped their ideas of how the US got counterinsurgency wrong in Vietnam.[43]

Two other points are worth emphasizing. First, the manual was not just a 'new' doctrine but was also a public relations exercise designed to demonstrate that the US military had a more thoughtful, humane, and culturally sensitive approach to the wars in Afghanistan and Iraq.[44] In an attempt to appeal to liberal angst over the war in Iraq, Sewell's introduction questions the rightness of the Iraq War even as it argued that the application of this counterinsurgency approach in Iraq needed to be actively supported by liberals.[45] One role of the new doctrine was to address a core challenge of counterinsurgency: the need to maintain domestic support for a protracted war.

Second, the shift to counterinsurgency was not simply a Bush-era Republican initiative. Indeed, the 'new' counterinsurgency doctrine arose not out of adherence to Bush's neoconservative commitments but, rather, a critique of their failures in Iraq and Afghanistan.[46] Emphasizing its connection to liberals, one conservative critic wrote that the new counterinsurgency doctrine was "warfare for northeastern graduate students—complex, blended with politics, designed to build countries rather than destroy them, and fashioned to minimize violence. It was a doctrine with particular appeal to people who would never own a gun."[47]

COUNTERINSURGENCY DOCTRINE, COLONIAL WARFARE, AND PACIFICATION

For military strategists, counterinsurgency is a specific kind of warfare, distinct from 'regular warfare' but also more than simply insurgency's mirror image.[48] Its origins lie in the practices of European colonial warfare, but its contemporary iterations have largely focused on the problem posed by Mao's development of a particular form of insurgency known as People's Protracted War. Indeed it is Galula's experience of the Red Army in China that motivated his preoccupation with counterinsurgency.[49] What distinguishes Mao's conception of 'People's War' is its understanding of the political nature of insurgency and its emphasis on political mobilization as the key to winning.[50] According to Thomas Hammes, Mao recognized and applied two critical concepts in the Chinese Revolution: "The first was that political power was the essential force in an armed conflict. The second was that political power could be used to change the correlation of forces so that insurgent conventional forces could conduct the final offensive to overthrow the government."[51] This led to a strategy of building support among the population and an insistence that military practices remain consistent with a 'spirit of unity' with the people.[52] There is much to say about Mao's approach to insurgency and its influence, but for now it is worth emphasizing that for counterinsurgency proponents, Mao's approach encapsulated their challenge. According to John Shy and Thomas Collier, "However much we may seek it elsewhere, the basic text for ideas about revolutionary war is in the writings of Mao Tse-Tung."[53]

American counterinsurgency doctrine describes insurgency and counterinsurgency as a subset of irregular warfare. Specifically:

> Insurgency [is] an organized movement aimed at the overthrow of a constituted government through the use of subversion and armed conflict. Stated another way, an insurgency is an organized protracted politico-military struggle designed to weaken the control and legitimacy of an established government, occupying power or other political authority while increasing insurgent control. Counterinsurgency is military, paramilitary, political, economic, psychological and civic action taken by a government to defeat insurgency.[54]

This definition highlights a few features that are distinctive about counterinsurgency in comparison to conventional warfare. First, if conventional war is concerned with destroying an opposing nation state(s) capacity to fight, the goal of counterinsurgency is preventing the overthrow of a 'legitimate' government ('legitimate' from the perspective of American interests, regardless of whether it is an established government or occupation) by insurgents. Quite often, as the US found in Afghanistan and Iraq, this 'legitimate' government must not only be buttressed by American aid, advisors, and boots on the ground, it must also be constructed. In both Iraq and Afghanistan, the US had to create states from the ground up. It is because of this that counterinsurgency become a 'war to build.'

'Legitimacy' is the core goal for counterinsurgency. According to the manual, the "primary objective … is to foster development of and effective governance by a legitimate government."[55] But this idea of legitimacy contains some ambiguities. For whom must the government be legitimate? Typically, counterinsurgents view this struggle for legitimacy as 'a battle for hearts and minds.' Peter Mansoor suggests that "The people must be convinced that support for the legitimate governing authority is preferable to support for the insurgent cause,"[56] a point which suggests that legitimacy has already been determined, whether the people like it or not. A second ambiguity is the meaning of legitimate government. Sometimes democracy is presented as a key feature of legitimacy but counterinsurgents are quick to suggest that democracy cannot be rushed and that there are other foundations for legitimacy in these less liberal societies.[57] Instead, counterinsurgents emphasise stability and the rule of law. According to the manual, "A government's respect for pre-existing and impersonal legal rules can provide the key to gaining it widespread, enduring societal support. Such government respect for rules—ideally ones recorded in a constitution and in laws adopted through a credible, democratic process—is the essence of the rule of law."[58] According to Kalev Sepp, "[i]t is in the national interest of all law-abiding countries to be able to fight insurgencies. If nations are indeed dedicated to the international rule of law and global stability, their governments and armed forces must be able to conduct or support counterinsurgency, counterterrorism, and stability operations"[59] Here, legitimacy, the practice of counterinsurgency, and a commitment to the rule of law become intrinsically bound together so that the violence of counterinsurgency is both a condition and a consequence of the promise of 'peace' that the rule of law is supposed to provide. This rule of law evokes an abstraction that substitutes for the conditions of stability and peace. Neocleous has noted how, beginning in the nineteenth century, international law, which contains a particular commitment to a bourgeois political economy, has been bound up with a division of the world into civilized and barbaric nations. The former was governed by the Law; the latter would necessarily be policed by the former for the sake of civilization.[60] What Sepp, and American national security doctrine bring into the definition of legitimacy implicitly is a particular bourgeois law constructed around facilitating the conditions of capital accumulation.

The goal of legitimacy shifts the focus from an insurgent enemy to the population in which the insurgent is embedded. This population-centric rather than enemy-centric emphasis is a crucial distinction and central to the 'new' counterinsurgency. It is only secondarily about killing the enemy, and far more about winning the population over to the legitimacy of the government or occupation authority. According to Galula, "The population represents this new ground. If the insurgent manages to dissociate the population from the counterinsurgent, to control it physically, to get its active support, he will win the war because, in the final analysis, the exercise of political power depends on the tacit or explicit agreement of the population or, at worst, on its submissiveness."[61] Whereas conventional war is about destroying enemy's capacity to fight, war in the service of politics, population-centric counterinsurgency is much more explicitly politics as war, or a war for the 'hearts and minds' of the occupied population. It is worth asking what the population being asked to participate in? The answer to that is of course a new order built around a 'legitimate' authority, legitimate because of its commitment to the rule of law, stability, and the market.

This population-centric counterinsurgency creates new demands on the military. Etienne de Durand, writing on the French experience of colonial warfare explains, borrowing from French colonial general Hubert Lyautey:

> *pacification*, must rely on a constant and subtle mix (*dosage*) of both coercion and consent. The former can be obtained through shows of force, capture of rebels, strict control of the population; the latter relies on the cooptation of local elites ("govern with the mandarin and not against the mandarin"), social works spreading the material advantages of modern civilization, and even missionaries. As "political action is by far what matters most," the country and its inhabitants should be handled with tact and consideration as far as possible, and destruction considered as a last resort.[62]

Saberi's account of Lyautey in this book highlights the ways in which his colonial warfare blended violence with urban design and social reform as one key element of this mix of coercion and consent. Counterinsurgency is characterized not just by the use of force but by a range of other 'civil' actions undertaken by the military as well as other state institutions (although the military is key due to a lack of capacity of the other actors).[63] The central role of these other kinds of actions is that they restore the 'legitimacy' of the established government through the construction of a new society. As Sewell explains, central to counterinsurgency are various 'nation-building' 'nonkinetic activities' such as providing 'electricity, jobs, and a functioning judicial system.'[64] For Mansoor, since "the population is the decisive terrain," victory requires the counterinsurgent convince "the people that a better life lies ahead."[65] Consequently contemporary counterinsurgency is about much more than just violence. It is also about 'nation-building' or 'development' that is consistent with the aims of the counterinsurgent.

Here, counterinsurgency becomes a 'war to build, not just destroy.' The provision of sewage, water, electricity, academics, trash, medical aid, and security (SWEAT-MS) is essential.[66] But this is not just charity. Mansoor suggests counterinsurgency must 'remember the first rule of economics: anything free will be overused.'[67] The price is that the people must be forced to 'make an active choice in favor of supporting the legitimate governing authority.'[68] Mansoor lets Kilcullen explain:

> The gratitude theory–'be nice to the people, meet their needs and they will feel grateful and stop supporting the insurgents'–does not work. ... the choice theory–enable (persuade, coerce, coopt) the population to make

an irrevocable choice to support COIN forces/government usually works better.[69] This 'choice,' of course, is a choice to participate in the legitimacy of the counterinsurgent state and its 'good works.' But as is clear, this moment of "construction of social works" and of "meet[ing] the needs of the population" is also one of violence, or at least the threat of it, if the people do not participate in the right way.

In applying the doctrine in a particular area, counterinsurgents use a Clear—Hold—Build approach. Clear operations remove insurgents from an area. Holding operations set up the infrastructure that embeds and sustains the counterinsurgent forces among the population. Building operations engage in the development necessary to build civil infrastructure, meet the needs of the people, and in so doing, permanently separate the insurgent from the population by 'legitimizing' the government.[70] This is modeled on the 'oil stain method' used in French colonial warfare. Rid, paraphrasing Lyautey, explains:

> Attacking an insurgency's foundation, or its soil, meant to turn the population into our foremost helper, … Instead of operating against the enemy's resources, markets were created and trade encouraged; instead of cutting communications, streets, rails and telegraph lines were built; instead of burning villages, settlements were protected and fortified; instead of killing the population, neutrals were trained and armed. … the objective was to grow and extend the secured and economically active zones from the center to the periphery, like oil spreads out on water.[71]

Clear—Hold—Build operates on a similar premise in setting up controlled zones and then incorporating and pacifying more territory and people.

According to the manual, the primary tasks of Clear—Hold—Build operations are to:

- Provide continuous security to the local population.
- Eliminate the insurgent presence.
- Reinforce political primacy.
- Enforce the Rule of Law
- Rebuild local [host nation] institutions.[72]

Apart from the direct violence necessary to clear and to hold the area, through each stage this approach entails two key components. The first is 'development' and the creation of 'legitimate' institutions.[73] The second is to ensure that the population is under the 'firm control' of the counterinsurgent through a variety of disciplinary measures. These include a complete and systematic census to determine the different social relationships, the issuing of identification cards, curfews, checkpoints, and other restrictions on the movement of people.[74] It's worth emphasizing two points out of this. First, that counterinsurgent victory requires turning "the population into our foremost helper," a point that Stuart Schrader emphasizes elsewhere in this book. That is, the population must internalize the commitment of the counterinsurgent to the Rule of Law and the goals of development as constructed by the counterinsurgent. Second, much like Mao's three stages of guerilla warfare,[75] the elements of Clear—Hold—Build, are fluid and entwined. Violence, political work, lawmaking and institution-building are present in each moment. Hence, as Schrader, notes, it was no stretch for American counterinsurgents to invert the French approach.

Its focus on reconstructing society makes counterinsurgency nothing less than a *total war*. In his account of the French wars in Vietnam and Algeria, Peter Paret noted that one of the explanations that French officers gave for their failure was their opponent's recognition that this was a total war and the French government and civilian's unwillingness to do the same.[76] This idea would later shape American explanations for their failure in Vietnam and their solution for Iraq and Afghanistan. There are a couple of ways that this occurs, which build directly from the French experience.

First counterinsurgency requires a 'Unity of Effort.' The need to construct political, social, and economic institutions entails both a larger role for civilian actors and consistency and coordination across the political, military, and humanitarian functions of counterinsurgency. While the military is meant to play a role in development, this role is temporary. Drawing from the French colonial warfare tradition, Galula writes: "The soldier must then be prepared to become ... a social worker, a civil engineer, a schoolteacher, a nurse, a boy scout. But only for as long as he cannot be replaced, for it's better to entrust civilian tasks to civilians."[77] In the long term, as Sewell explains, counterinsurgency requires civilians at "every level of operations."[78] Consequently, if the US does what is necessary, then "civilian actors and agencies would become centrally engaged in the field alongside combat forces and the risks and costs of counterinsurgency would be spread across the US government."[79] The manual devotes a chapter to this, focusing on the role of civilian actors (including government, non-governmental, international actors as well as the private sector) and on the need for coordination of civilian and military actors.[80] Responsibility for success extends beyond the military to the intervening state and society. In the literature, the necessity of this unity of effort is built from the colonial experience and the necessary role of the civilians in the form of a colonial bureaucracy and civilian colonial experts.[81]

Second, societies engaged in counterinsurgency need to prepare for this long-term commitment. According to the manual, "Insurgencies are protracted by nature, and COIN operations always demand considerable expenditures of time and resources."[82] US forces must convince populations of their steadfastness, long-term headquarters and support structures need to be established, and some support for host nation institutions usually remains long after the military presence has been reduced.[83] Sewell puts it this way: "These demands are the essence of the doctrine. Will Americans supply greater concentrations of forces, accept higher casualties, fund serious nation-building, and stay many long years to conduct counterinsurgency by the book."[84] For her this is the route the US must take if it wants both success and 'decency.'[85]

In seeking to reorganize the American military for counterinsurgency, they also want a different kind of soldier, constructed on a colonial past. Rid, paraphrasing Lyautey, notes that 'those who fought on a daily basis in remote lands against ferocious enemies and the equally relentless elements, those who went through that "rough school," were "neither *military* nor *civilian* any more, but simply *colonial*."[86] For De Durand, the key lesson to be drawn from French colonial experience is that "it is by going native, by speaking the language(s), and by developing an intimate knowledge of the human terrain that European officers were able to manipulate local politics and social dynamics effectively."[87] Robert Scales writes that "Soldiers must gain the ability to move comfortably among alien cultures, to establish trust and cement relationships that can be exploited in battle."[88] Consequently Scales, and other counterinsurgents emphasize the different skills associated with language, cultural knowledge, and development that US soldiers will need to learn.[89]

Counterinsurgency also demands the development of capacities to understand the so-called human terrain, that is the "population and society in the operational environment."[90] The definition of military intelligence is broadened to include a whole range of social, cultural, and linguistic concerns.[91] In *Orientalism,* Edward Said noted that "Lurking everywhere behind the pacification of the subject race is imperial might, more effective for its refined understanding and infrequent use than for its soldiers, brutal tax gatherers, and incontinent force. In a word, the Empire must be wise."[92] This is a lesson taken to heart by contemporary counterinsurgents, who see knowledge or a 'conflict ethnography' as crucial to the struggle for the hearts and minds of the population.[93]

Social and cultural knowledge is deemed crucial to counterinsurgency, and as has been well documented now, the US military introduced Human Terrain Teams of social scientists to join platoons on the ground in Iraq and Afghanistan in collecting this information.[94] As the 'the graduate level of war' or a 'World War IV' that will be won by the social scientists, counterinsurgency is intended to be multidisciplinary and include historians, political scientists, strategists, area specialists, anthropologists, sociologists.[95] The manual emphasizes that "Counterinsurgents must understand the environment," meaning the society and culture of the people in the area of operation and including:

- Organization of key groups in the society.
- Relationships and tensions among groups.

- Ideologies and narratives that resonate with groups.
- Values of groups (including tribes), interests, and motivations.
- Means by which groups (including tribes) communicate.
- The society's leadership system.[96]

Mansoor writes that "the only way to protect a people effectively is to live among them. Leaders must understand native culture and study local history in order to determine what motivates the people."[97] This knowledge is crucial precisely because the population is now the 'terrain' of struggle. It enables the counterinsurgent to determine how and to what extent one can manage, mobilize, and repress the population in the construction of 'legitimacy.'

Counterinsurgents also seek to reconstruct the security forces of their local allies through training and advisory missions and the creation of local proxies. As Nagl explains, a dilemma of counterinsurgency is that the counterinsurgent must be strong everywhere, and all the time, whereas the insurgent only rarely needs to be strong, and in a few places. Counterinsurgency, for its American proponents, requires a minimum 20:1000 counterinsurgent: civilian ratio. This makes it prohibitive without the aid of local forces.[98] Scales writes, "WW4 will be manpower-intensive. The US cannot hope to field enough soldiers to be effective [and] surrogates are needed to help us fight our wars."[99] Consequently, the US needs to make significant use of local forces and reorganize its own forces to facilitate the training of "loyal indigenous armies."[100] For Nagl, this means that US conventional-purpose forces must be prepared for advisory roles as long as the US "remains engaged in an advisory role in the Long War." He quotes Lt. General James Dubik: "The conventional forces of the US Army will have an enduring requirement to build the security forces and security ministries of other countries. This requirement is consequently not an aberration, unique to Iraq and Afghanistan."[101] The role of local forces is also that they meet core needs of counterinsurgency. According to Nagl, "They know the terrain, both physical and human, and generally speak the language. They understand the social networks that comprise the society and how they are interrelated."[102] A training and advisory role has been further emphasized in the 2014 update of the manual as a way to get around both the political and economic costs of counterinsurgency.[103] This emphasis is not at all a break from counterinsurgency. It is simply a shifting of the responsibility for counterinsurgency and its attendant obligations to enforce the rule of law, maintain stability and protect the new 'legitimate' institutions onto American backed and trained militaries of the states they have supported or constructed.

In the above section, two things were emphasized. First, that colonial warfare practices are deeply embedded in counterinsurgency doctrine. Second, this doctrine was conceived as 'a war to build rather than destroy.' In counterinsurgency, the aim is to create a particular kind of stability and 'legitimate' order through the creation of state institutions, the establishment of the rule of law, and the fostering of economic development. Violence is integral to this, but so too is a set of other techniques that borrow from the colonial period and which promise to moderate violence through other kinds of civil action. It is because of this that counterinsurgency is considered a form of warfare where political factors are primary.[104] But it is important to recognize that another reason counterinsurgency is a more political form of war is because it is a social war to secure the conditions of capital accumulation.

COUNTERINSURGENCY AS GLOBAL SOCIAL WAR

Neocleous reminds us that for Marx and Engels, the history of primitive accumulation was also a history of social war.[105] People were dispossessed and a proletariat was constructed, a process that occurred alongside the domination of colonial subjects. The state played a central role in both their creation and regulation to create the conditions of accumulation. But, as Harvey and Neocleous have noted, this was not simply an historical process. Rather, as Neocleous puts it, it is "a *permanent feature of accumulation* and an *integral component of class war*" [emphasis in original].[106]

Globalization, while a consequence of capital's drive to accumulate, also confronts capital with obstacles. Here I just want to focus on two examples in the Marxist literature on imperialism. For Ellen Wood, this "surplus imperialism" that concentrates military violence directed at no definable target only makes sense as a response to the "global state system and its contradictory dynamics." By this she means that capital, especially in this globalized moment, needs states to create the conditions that capital needs, but it also needs the right states, and it needs to create or discipline them where they don't exist.[107] Similarly, for Samir Amin, "Peaceful management of the new imperialist system ... becomes less and less of a possibility, as political violence, and hence military intervention, serves functions essential to a project that misleadingly describes itself as "liberal.""[108] What these highlight is the way that the conditions that will enable the internationalization of capital are a core pillar of national security for the leading capitalist states and specifically for the United States. Counterinsurgency as an element of that violence is absent but not inconsistent with these accounts.

This account of imperialism has been internalized by contemporary counterinsurgents who understand contemporary insurgency to be distinctive precisely because it is global in scope. Its global character in Kilcullen's view is thought to be a product of the spaces that globalization has left behind, or that refuse to participate in global economic processes as well as the new tools and techniques that globalization has enabled.[109] For Nagl "[f]inal victory in today's fight depends upon the integration of the nations in the Arc of Instability into the globalized world's economic and political system."[110] For its proponents, counterinsurgency is necessary precisely so that integration occurs to bring that 'Arc of Instability' into the orbit of global processes of accumulation. To do so it must, as the liberal apologists euphemize, civilize the barbarians. In this sense, it can be conceived as a global social war.

It is worth concluding with a comment on the future of this counterinsurgency doctrine. In the years following the 'success' of the surge in Iraq, the sheen has worn off of the 'new' counterinsurgency. The insurgency in Afghanistan has not abated, and ISIL has emerged in Iraq and Syria to challenge US interests there. Counterinsurgency's key proponent, Petraeus, having been appointed director of the CIA in September 2011, left in disgrace by November 2012. The 'national security interests' of the US in the region are apparently undermined by the corrupt regimes that they have constructed and backed, and American society is weary of American casualties abroad. Further, even if counterinsurgency involves all sorts of 'nonkinetic' action in the form of 'development' or nation-building, the lie of a gentler approach has been repeatedly demonstrated by the violence that comes with it. The world that the counterinsurgents had sought to construct in Iraq and Afghanistan is in shambles and the 'dream of pacification' that visited a nightmare upon the people to be pacified has become one for its practitioners as well.

But it is too early to speak, as some have, of the death of counterinsurgency doctrine.[111] Certainly the military has had to re-evaluate its counterinsurgency approach. The newest iteration of the manual has shifted emphasis further to training and advisory missions to avoid the costs of more direct US participation. But its understanding of the problem and counterinsurgent methods remain largely the same. Nor is failure or even the inconsistency between counterinsurgency claims and its reality on the ground reason to see its end. In a debate on its future, Nagl argued that counterinsurgency could not be dead because insurgency was not yet dead.[112] Indeed, as has been suggested, the emphasis on counterinsurgency is driven by capitalism's drive to accumulate and its need to construct the space and the populations that enable that. Pacification, as Neocleous has noted, is a dream, but one that capital's agents seek to refine, not escape.

I began the chapter quoting Karl Liebknecht's account of the importance of the history of militarism to capitalism. For Liebknecht, this history of militarism would reveal "the most deeply hidden and delicate root-fibres of capitalism." Drawing from this idea, I have argued that recent developments in American military doctrine are important to understanding imperialism, and that a useful lens to do so, at least in regards to counterinsurgency, is through this critical reading of pacification. In emphasizing a war to build as well as destroy, pacification both links counterinsurgency back to its colonial past and to capital's dream of a docile world available for its accumulation.

NOTES

1. I would like to thank Hülya Arik, Parastou Saberi, Jerome Klassen, Mark Neocleous, Tyler Wall, Stuart Schrader, and Joshua Moufawad-Paul as well as the anonymous reviewers for their feedback and thoughts on the writing of this chapter.
2. Karl Liebknecht, *Militarism* (Toronto: William Briggs, 1917), 2.
3. John Nagl, *Learning to Eat Soup With a Knife: Counterinsurgency Lessons from Malaya and Vietnam* (Chicago: University of Chicago Press, 2005), xvi.
4. Dan Bolger, quoted in Nagl, *Learning to Eat Soup With a Knife*, 222-223.
5. Sarah Sewell, "Introduction to the University of Chicago Press Edition" in *The US Army/Marine Corps Counterinsurgency Field Manual: US Army Field Manual no. 3-24: Marine Corps warfighting publication no. 3-33.5*. eds. United States et al. (Chicago: University of Chicago Press, 2007), xlii-xliii.
6. Michael Ignatieff, *Empire Lite: Nation Building in Bosnia, Kosovo, Afghanistan* (Toronto: Penguin Canada, 2003).
7. Sewell, "Introduction to the University of Chicago Press Edition," xxxiv.
8. Montgomery McFate, "Anthropology and Counterinsurgency: The Strange Story of Their Curious Relationship," *Military Review* (March-April 2005), 24.
9. McFate, "Anthropology and Counterinsurgency," 26.
10. Nagl, *Learning to Eat Soup With a Knife*.
11. For an account of Lyautey's approach, see Parastou Saberi's chapter in this book.
12. Karl Marx, "The Communist Manifesto," in *The Portable Karl Marx*, ed. Eugene Kamenka (Markham: Penguin Brooks, 1983), 208.
13. A useful, if dated, review of Marxist approaches to imperialism is Anthony Brewer, *Marxist Theories of Imperialism: A Critical Survey*, 2nd ed. (London: Routledge, 1990).
14. A good brief review comparing Marxist and more mainstream realist and neoliberal accounts of imperialism can be found in Jerome Klassen's recent *Joining Empire: The Political Economy of the New Canadian Foreign Policy*, (Toronto: University of Toronto Press, 2014). See chapter 1, "Understanding Empire: Theories of International Political Economy."
15. Among many others, see Klassen, *Joining Empire*; Leo Panitch and Sam Gindin, eds., *Socialist Register, 2004* (London: Merlin Press, 2003); David Harvey, *The New Imperialism* (New York: Oxford University Press, 2005); Samir Amin, *Obsolescent Capitalism: Contemporary Politics and Global Disorder* (London: Zed Books, 2003); Ellen Wood, *Empire of Capital* (London: Verso Books, 2005).
16. Harvey, *The New Imperialism*.
17. Harvey, *The New Imperialism*, 138-143, 145-149.
18. Harvey, *The New Imperialism*, 195-199, 145-146. Klassen, *Joining Empire*, 63.
19. Mark Neocleous, "The Dream of Pacification: Accumulation, Class War, and the Hunt," *Socialist Studies*, 9, no. 2, (2013): 8.
20. Neocleous, "The Dream of Pacification," 11.
21. Neocleous, "The Dream of Pacification: Accumulation, Class War, and the Hunt," 8.
22. Neocleous, "The Dream of Pacification: Accumulation, Class War, and the Hunt," 18.
23. Neocleous, "The Dream of Pacification: Accumulation, Class War, and the Hunt," 9.
24. Mark Neocleous, *War Power, Police Power* (Edinburgh: Edinburgh University Press, 2014),135-136.
25. Naomi Klein, *The Shock Doctrine: The Rise of Disaster Capitalism* (Toronto: Vintage Canada Edition, 2008) and Pratap Chaterjee, *Iraq Inc: A Profitable Occupation* (2004).
26. Klein, *The Shock Doctrine: The Rise of Disaster Capitalism*. See also Rajiv Chandresakaran, *Imperial Life in the Emerald City: Inside Iraq's Green Zone* (New York: Alfred A. Knopf, 2006) and David Stokes, "The War Gamble: Understanding US Interests in Iraq," *Globalizations*, 6, no. 1, 2009.
27. Mike Skinner, "The Empire of Capital and the Latest Inning of the Great Game" in *Empire's Ally: Canada and the War in Afghanistan*, eds. Jerome Klassen and Greg Albo (Toronto: University of Toronto Press, 2013); David Stokes, "The War Gamble."
28. Skinner, "The Empire of Capital," 120-122.
29. Niall Ferguson, quoted in Klassen, *Joining Empire*, 33-34.

30 See Klassen, *Joining Empire*, 34-35. An explicit commitment to an open international economic system and a rules-based international order has been a reoccurring theme of American National Security Doctrine. See, most recently, United States, *United States National Security Strategy 2015*, February 2015, accessed July 31st 2015. https://www.whitehouse.gov/sites/default/files/docs/2015_national_security_strategy.pdf.
31 Klassen, *Joining Empire, 62*. See also Greg Albo "The Old and New Economics of Imperialism" in *Socialist Register 2004: The New Imperial Challenge*; Adam Hanieh, "A 'Single War': The Political Economy of Intervention in the Middle East and Central Asia" in *Empire's Ally: Canada and the War in Afghanistan*, eds. Jerome Klassen and Greg Albo (Toronto: University of Toronto Press, 2013).
32 See Klassen, *Joining Empire*, 41.
33 Thomas Keaney and Thomas Rid, "Understanding Counterinsurgency," in *Understanding Counterinsurgency: Doctrine, Operations, and Challenges*, eds. Thomas Keaney and Thomas Rid (New York: Routledge, 2010). For a an account of the manual's development and those involved see Conrad Crane, "United States," in *Understanding Counterinsurgency: Doctrine, Operations, and Challenges*, eds. Thomas Keaney and Thomas Rid (New York: Routledge, 2010).
34 Andrew Krepenivich, *The Army and Vietnam* (Baltimore: John Hopkins University Press, 1986), 4-7; Nagl, *Learning to Eat Soup With a Knife*, 115-116; McFate, "Anthropology and Counterinsurgency," 26-27.
35 Nagl, "Forward to the University of Chicago Press Edition," in *US*, eds., United States, John Nagl, David Petraeus, and James Amos (Chicago: University of Chicago Press, 2007), xiv.
36 Kalev Sepp, "Special Forces," in *Understanding Counterinsurgency: Doctrine, Operations, and Challenges*, eds. Thomas Keaney and Thomas Rid (New York: Routledge, 2010), 128.
37 David Ucko, *The New Counterinsurgency Era: Transforming the US Military for Modern Wars* (Washington D.C.: Georgetown University Press, 2009), 61.
38 Ucko, *The New Counterinsurgency Era*, 47-54. Also see Thomas Hammes, *The Sling and the Stone: On War in the 21st Century* (St. Paul: Zenith Press, 2004), 5-10.
39 Gian Gentile, "The Army's Learning and Adapting Dogma," in *The National Interest*, September 16 2013, accessed July 31, 2015, http://nationalinterest.org/commentary/the-armys-learning-adapting-dogma-9070. Gentile also provides a useful summary of the pro-counterinsurgency narrative.
40 Thomas Rid, "Nineteenth Century Origins of COIN Doctrine," *The Journal of Strategic Studies*, 33, no. 5, 730.
41 Ann Marlowe, *David Galula: His Intellectual Life and Context* (Carlisle: Strategic Studies Institute, 2010), 48.
42 Marlowe, *David Galula: His Intellectual Life and Context*.
43 Marlowe, *David Galula: His Intellectual Life and Context*, 10. Krepenivich, *The Army and Vietnam*, 277, endnote 5. This emphasis on Galula has also met skepticism from Rid and others who see Galula as overrated and as drawing from the French colonial warfare tradition rather than as original thinker. Rid, "Nineteenth Century Origins of COIN Doctrine," 728, 731.
44 Robert Gonzales, *American Counterinsurgency: Human Science and the Human Terrain* (Chicago: Prickly Paradigm Press, 2009), 10, 13-18.
45 Sewell, "Introduction to the University of Chicago Press Edition."
46 Ucko, *The New Counterinsurgency Era*, 51-52, 60-61.
47 Charles Dunlap Jr., "Airpower," in *Understanding Counterinsurgency: Doctrine, Operations, and Challenge*, eds. Thomas Keaney and Thomas Rid (New York: Routledge, 2010), 102. Dunlap is quoting Steve Coll, "The General's Dilemma: David Petraeus, the Pressures of Politics, and the Road Out of Iraq," in *New Yorker*, September 8, 2008, www.newyorker.com/reporting/2008/09/08/080908fa_fact_coll.
48 United States, John Nagl, David Petraeus, and James Amos, *The US Army/Marine Corps Counterinsurgency Field Manual: US Army Field Manual no. 3-24: Marine Corps warfighting publication no. 3-33.5*. (Chicago: University of Chicago Press, 2007), 2.
49 Marlowe, *David Galula: His Intellectual Life and Context*, 28.
50 Mao Zedong (1937), *On Guerrilla Warfare*, trans. Samuel Griffith (Garden City: Anchor Books, 1978); United States et al., *Counterinsurgency Field Manual*, pp. 11-14; Hammes, *The Sling and the Stone*, 51-52.
51 Hammes, *The Sling and the Stone*, 52-53.
52 Nagl, *Learning to Eat Soup With a Knife*, 22.

53 John Shy and Thomas Collier, "Revolutionary War" in *Makers of Modern Strategy: From Machiavelli to the Nuclear Age,* ed. Peter Paret (Princeton: Princeton University Press, 1986); quoted in Nagl, *Learning to Eat Soup With a Knife,* 19.
54 United States, *Counterinsurgency Field Manual,* 2.
55 United States, *Counterinsurgency Field Manual,* 37.
56 Peter Mansoor, "The Army," in *Understanding Counterinsurgency: Doctrine, Operations, and Challenges,* eds. Thomas Keaney and Thomas Rid (New York: Routledge, 2010), 82.
57 Octavian Manea and John Nagl, "COIN is not Dead. An interview with John Nagl," *Small Wars Journal* (February 6, 2012), http:/smallwarsjournal.com/jrnl/art/coin-is-not-dead.
58 United States, *Counterinsurgency Field Manual,* 39.
59 Sepp, "Special Forces," 128.
60 Neocleous, *War Power, Police Power,* 125-127.
61 Galula, quoted in Mansoor, "The Army," 78-79.
62 Etienne de Durand, "France," in *Understanding Counterinsurgency: Doctrine, Operations, and Challenges,* eds. Thomas Keaney and Thomas Rid (New York: Routledge, 2010), 14.
63 Nadia Schadlow, "Governance," in *Understanding Counterinsurgency: Doctrine, Operations, and Challenges,* eds. Thomas Keaney and Thomas Rid (New York: Routledge, 2010), 183.
64 Sewell, "Introduction to the University of Chicago Press Edition," xxx.
65 Mansoor, "The Army," 82.
66 Ibid., 82-83.
67 Ibid., 83.
68 Ibid.
69 Kilcullen, quoted in Mansoor, "The Army," 83-84.
70 Mansoor, "The Army," 75.
71 Rid, "Nineteenth Century Origins of COIN Doctrine," 752.
72 United States, *Counterinsurgency Field Manual,* 174-175.
73 United States, *Counterinsurgency Field Manual,* 181-182.
74 United States, *Counterinsurgency Field Manual,* 180-181.
75 Mao Zedong, *Mao on Warfare: Guerrilla Warfare, On Protracted War, and Other Martial Writings* (New York: CN Times Books, 2013), 170.
76 Peter Paret, *French Revolutionary Warfare from Indochina to Algeria: The Analysis of a Political and Material Doctrine* (New York: F.A. Praeger, 1964), 26-30.
77 United States, *Counterinsurgency Field Manual,* 68.
78 Sewell, "Introduction to the University of Chicago Press Edition," xxix.
79 Sewell, "Introduction to the University of Chicago Press Edition," xxxii.
80 United States, *Counterinsurgency Field Manual,* Chapter 2. See also Schadlow, "Governance."
81 See for example, Nagl, *Learning to Eat Soup With a Knife,* or McFate, "Anthropology and Counterinsurgency."
82 United States, *Counterinsurgency Field Manual,* 43.
83 United States, *Counterinsurgency Field Manual,* 43-44.
84 Sewell, "Introduction to the University of Chicago Press Edition," xxxviii-xxxix.
85 Sewell, "Introduction to the University of Chicago Press Edition," xxxviii-xxxix.
86 Rid, "Nineteenth Century Origins of COIN Doctrine," 754.
87 de Durand, "France,"14-15.
88 Robert Scales, "Clausewitz and World War IV," *Armed Forces Journal* (July 1, 2006), accessed July 31, 2015, http://www.armedforcesjournal.com/clausewitz-and-world-war-iv/.
89 In addition to Scales, see for example Sepp, "Special Forces," or Paul Yingling, "A Failure of Generalship," *Armed Forces Journal* (May 1, 2007), accessed July 31, 2015, http://www.armedforcesjournal.com/a-failure-in-generalship.
90 Jacob Kipp, Lester Grau, Karl Prinslow, and Don Smith, "The Human Terrain System: A CORDS for the 21st Century," *Military Review* (September-October, 2006). See Endnote 2.
91 See for example United States et al., *Counterinsurgency Field Manual,* Chapter 3, or David Kilcullen, "Intelligence" in *Understanding Counterinsurgency: Doctrine, Operations, and Challenges,* eds. Thomas Keaney and Thomas Rid (New York: Routledge, 2010).
92 Edward Said, *Orientalism* (New York: Pantheon, 1978), 36-37.
93 David Kilcullen, "Intelligence," 154-155.
94 For a critique, see Gonzalez, *American Counterinsurgency.*
95 Scales, "Clausewitz and World War IV"; Rid and Kearney, "Understanding COIN," 2.
96 United States, *Counterinsurgency Field Manual,* 40.
97 Mansoor, "The Army," 79.

98 Nagl, "Local Security Forces," in *Understanding Counterinsurgency: Doctrine, Operations, and Challenges*, eds. Thomas Keaney and Thomas Rid (New York: Routledge, 2010), 160-170, 161-162; United States, *Counterinsurgency Field Manual*, 23.
99 Scales, "Clausewitz and World War IV."
100 Scales, "Clausewitz and World War IV."
101 Nagl, "Local Security Forces," 168.
102 Nagl, "Local Security Forces," 161.
103 Walter Ladwig III, "The New FM 3-24: What Happens When the Host Nation is the Problem?' *Defense in Depth*, ed. Janine Davidson (Council of Foreign Relations), June 10, 2014. Accessed July 31, 2015, http://blogs.cfr.org/davidson/2014/06/10/the-new-fm-3-24-what-happens-when-the-host-nation-is-the-problem/; The US Army/Marine Corps, *Insurgencies and Countering Insurgencies: US Army Counterinsurgency Field Manual no. 3-24: Marine Corps warfighting publication no. 3-33.5*, 2014; David Ucko and Robert Egnell, "On Military Interventions: Options for Avoiding Counterinsurgencies," *Parameters*, 44, no. 1 (2014): 17.
104 United States, *Counterinsurgency Field Manual*, 2.
105 Neocleous, "The Dream of Pacification," 11-12.
106 Neocleous, *War Power, Police Power*, 85.
107 Wood, *Empire of Capital*, 154-157.
108 Amin, *Obsolescent Capitalism*, 76.
109 David Kilcullen, The *Accidental Guerrilla: Fighting Small Wars in the Midst of a Big one* (Oxford: Oxford University Press, 2009), 7-16.
110 Nagl, *Learning to Eat Soup With a Knife*, xvi.
111 Ryan Evans, "COIN is Dead, Long Live the New COIN," *Foreign Policy* (December 16, 2011), accessed July 31, 2015, http://foreignpolicy.com/2011/12/16/coin-is-dead-long-live-the-coin/.
112 A.A. Cohen, "Cage Match in a Corn Field: Gian Gentile wrestles J. Nagl on Counterinsurgency," *Foreign* Policy (May 30, 2013), accessed July 31, 2015, http://foreignpolicy.com/2013/05/30/cage-match-in-a-cornfield-g-gentile-wrestles-j-nagl-on-counterinsurgency.

'Going All Out For Shale'
Fracking as Primitive Accumulation

Will Jackson, Helen Monk, and Joanna Gilmore

Hydraulic fracturing, better known as 'fracking,' is now a central pillar of UK government strategy on 'energy security.' This is no surprise. The rapid rise of fracking in the UK, facilitated by current and previous governments, is based on an attempt to replicate the fracking boom in the US. Technological advances in the last 25 years enabled the exploitation of previously inaccessible shale gas reserves in the US, and this facilitated the rapid expansion of a fracking industry that from a relatively slow start in the late 1980s was able to grow by 45% a year between 2005 and 2010.[1] Expansion in supply in the US has outpaced all forecasts, and as a proportion of America's overall gas production, shale gas increased from 4% in 2005 to 24% in 2012.[2] The scale of this expansion has been, and continues to be, enabled by collaboration between state and capital producing the necessary regulatory environment and investment in infrastructure.

The experience in the US has given rise to a 'dash for gas' across Europe in which energy companies and governments have sought to move quickly to exploit unconventional shale reserves identified across the continent.[3] In the last decade the technological advancements developed in the US—specifically the merger of hydraulic fracturing and horizontal drilling techniques—have been exported around the globe.[4] In the UK, significant shale deposits have been identified[5] and exploratory drilling to explore these 'new frontiers' has been actively encouraged by government since 2007.[6]

However, at these frontiers the industry has met resistance in the UK, as it has elsewhere. In the UK, new coalitions of local opponents and more established climate and social justice groups have focussed on the risks of environmental degradation seemingly inherent in fracking, with campaigners pointing to the real environmental impacts already documented in the US.[7] Fracking involves the pumping of water, proppants (sand or similar manufactured granules) and assorted chemicals into the ground at high pressure[8] and the concerns of opponents have focussed on the potential for land, air and water pollution,[9] seismic instability,[10] and the broader issue of maintaining a reliance on carbon intensive fossil fuels in the face of global climate change.[11]

Opposition to fracking has centred on the fact that where fracking has been rolled out it has 'significantly altered physical environments, economic systems, community structures and human health.'[12] However, what the opposition has made most obvious is that fracking stands as a contemporary process of enclosure; to use David Harvey's term, "accumulation by dispossession."[13] For Timothé Feodoroff et al., fracking is first and foremost bound up in "the contemporary global land grab" reliant as it is on the capture of land, the large scale use of land, water, and capital, and the deepening of an extractive model of development.[14] As fracking develops across the globe, encouraged by state-capital alliances, opposition movements continue to grow.[15] These campaigns have started from a focus on the local impacts of fracking, but they are linked through their struggles against the further privatisation and depletion of the environmental commons.

Opposition to fracking is seemingly ubiquitous but the form it takes and its capacity for effective disruption is dependent on the political and legal context in which fracking develops. This chapter considers the issue of fracking—its general rationale, the resistance to it, and the policing of that resistance—through the lens of *pacification*. We argue that the police response to resistance movements in this context lays bare the process of pacification and demonstrates clearly both the destructive and productive dimensions of police power as expounded in the recent pacification literature.[16] Through a case study of the police response to anti-fracking protests at Barton Moss, Salford, UK between November 2013 and April 2014,[17] the analysis here illustrates that the production of 'docile bodies' inherent to the pacification process[18] is always at the heart of police responses to effective forms of protest. In particular, the chapter builds on the connection between pacification and *primitive accumulation*. Building on the connections made between pacification and primitive accumulation by Neocleous,[19] we want to suggest that fracking be understood as a form of primitive accumulation. In other words, and in contrast to much of the literature on fracking, this means treating fracking as a form of class war.[20]

COMMAND AND CONTROL

The first major show of force by anti-fracking protesters in the UK came in the Sussex village of Balcombe in summer 2013. The opposition—a combination of local opponents and more established environmental campaigners and anti-fracking activists—focussed their interventions on blockades of the drilling site and attempts to raise awareness about the existence of, and dangers associated with, fracking.[21] The protest was met with a significant joint police operation conducted by Sussex and Metropolitan police forces and this drew in media attention from around the country because of its status as the first major anti-fracking protest, but also partly due to the involvement and arrest of Green MP Caroline Lucas. The protest resulted in the company involved, Caudrilla, withdrawing (temporarily at least) its fracking application.[22]

Shortly after the events in Balcombe, the energy company IGas, specialists in onshore extraction of oil and gas, began exploratory drilling at Barton Moss, Salford in the northwest of England. Before the drilling started in November 2013, a protest camp was established at the site of the well by local residents, who were joined by a number of activists who had been involved at Balcombe. The camp was positioned along the approach road to the drill site and remained in place for the duration of the exploratory drilling until April 2014.

Over the period of protest, the camp gathered momentum and established itself as a community-led protection camp, sustained by local support and donations. Its residents, referring to themselves in many cases as 'protectors' rather than protesters, adopted several protest techniques, including the use of lock-ons[23] and blockades, but they relied most heavily on slow walking, or marching, in front of IGas convoys to disrupt and delay the drilling operation and to provide a visible and constant opposition to fracking in Salford. More broadly, the camp aimed to make the above interventions to highlight the universal hazards associated with fracking and the impact these could have on future generations, as well as drawing attention to the rapid development of the fracking industry in the UK.

These actions elicited a tough response from Greater Manchester Police (hereafter GMP) and an increasing number of Tactical Aid Unit (public order) officers, who responded to the protest with a substantial police presence at almost every march. The operation was codenamed Operation Geraldton. There were over 200 arrests—including the detention of children, pregnant, and elderly protesters, and the violent arrest of women—alongside many additional reports of police misconduct related to GMP's management of the protest[24]. The camp witnessed a notable broadening in focus following the intensification of the police presence and policing tactics. The right to protest, in the face of police violence, became a dual concern that underpinned many of the actions of camp members and local protesters. This also altered the role of the protesters who began to focus their protests on the police, establishing a second protest camp outside the headquarters of GMP. Camp residents and supporters established the 'Justice 4 Barton Moss' campaign in response to "aggressive, intimidating and violent policing by Greater Manchester Police."[25] This facilitated the development of links with a number of other justice campaigns in the UK focussed on policing and specifically deaths in police custody, as connections were made publicly by the groups between lethal police violence (including that by GMP) and the police response to protest.

The Chief Constable of GMP, Peter Fahy, defined the role of the police at Barton Moss as balancing the rights of protesters with those of IGas and publicly expressed his frustration at being "stuck in the middle."[26] The Chief Constable and other senior spokespeople at GMP sought at all times to explain their approach through the language of human rights, reiterating their respect for the right to "peaceful, legal protest" and consistently described their intervention at Barton Moss as purely reactive in response to the protesters' actions. Yet the policing at Barton Moss demonstrated that 'peaceful' protest in this sense only applies to that which is non-disruptive; protest that eschews all physical violence and aggression is not peaceful, and thus not permitted, if it remains disruptive. At Barton Moss, the protesters sought at great lengths to demonstrate a commitment to peaceful, non-violent protest but this included a commitment to non-violent direct action with the aim to disrupt the drilling operation. In this sense, the protesters transcended the boundaries of acceptable protest and their rights were not respected. This is not new or novel but it is important in understanding how the police and protesters presented the protest through competing narratives. These narratives revolved around opposing definitions of what constitutes a peaceful protest as well as competing rights claims as police sought to 'balance' the legal rights of protesters and IGas.[27]

The policing operation relied upon the regular use of Tactical Aid Unit (hereafter TAU) officers to limit the disruptive potential of the protest. These public order officers are officially available to provide "additional help to local officers with *unusual incidents,*"[28] but their deployment from the outset of Operation Geraldton was clearly planned and quickly normalised. Much of the violence experienced by protesters was a result of TAU interventions. The central role for TAU officers was to speed up the daily slow marches and reduce the time taken to get trucks in and out of the drilling site. As the protest developed, use of TAU officers became more frequent, and they often took the place of regular officers when they deemed that the pace of the march was too slow. The running battle between protesters and police revolved around the length of these marches; depending on who had the upper hand, the time taken to get trucks along the 800 metre long entry road ranged from several hours to as little as fifteen minutes. Towards the end of the drilling operation, the number of arrests and the reports of police brutality increased, leading the solicitor representing most of those arrested to state that the TAU officers appeared "out of control."[29] The various reports and online videos of the police violence at Barton Moss suggested to many that there had been a departure from 'normal' policing. We want to suggest an alternative way of thinking about this issue: that what was at stake in the policing was nothing less than the process of pacification.

Building on the work of Neocleous and Rigakos,[30] and in line with other contributions to this volume, pacification can be usefully reclaimed as a tool for a critical theory that seeks to understand the relationship between police and capital.[31] We want to suggest that this reclamation is essential to the project of seeking to understand the policing described above. Crucially, the history of pacification demonstrates that it is a *productive* process, never solely involved in simply suppressing resistance in imperial or domestic settings, but instead employed consciously in the construction of a new social order.[32]

In employing the concept of pacification to analyses of policing, we are able to render more visible the productive dimension to the exercise of police power and to understand that the production of liberal order has been, and continues to be, facilitated through the creation of the conditions for capitalist accumulation. In this sense pacification involves the production of 'docile bodies' ripe for economic exploitation and the policing of both colony and metropolis has always involved this 'creative' application of power. Pacification is thus understood best as a "political technology for organizing everyday life through the production and re-organization of the ideal citizen-subjects of capitalism."[33] A capitalist social order produces its ideal citizen-subjects—economically productive yet docile, disciplined, unable and unwilling to resist—by "shaping the behaviour of individuals, groups and classes, and thereby ordering the social relations of power around a particular regime of accumulation."[34] This is the productive dimension of police power: violence is employed not simply to supress but to produce, to build, to establish order.

Yet this ideal citizen-subject is just that, an *ideal*, a goal never fully achieved. The pacification process is always met with struggle in some shape or form; those who resist the violence of capitalist social relations *are* the disorder that must be eradicated. The conditions for capitalist accumulation and its ideal subjects neither appear organically nor are reproduced with universal consent; the exercise of state power as police power has been, and continues to be, essential to produce and sustain the inequality inherent in a capitalist social order. The response to struggle, defined invariably as *dis*order, is central to policing and in turn affects the form that policing takes; pacification is never a finished project—it is confronted by, shaped by, and ultimately limited by, struggle. In Rigakos and Ergul's terms, "pacification is the continuum of police violence upon which the fabrication of capitalist order is planned, enforced and resisted."[35]

Policing must therefore be understood as a project involving confrontation with 'disorderly' and 'disruptive' subjects and the creation of orderly, disciplined subjects conducive to the maintenance of a capitalist social order. The ideal citizen-subject is orderly in the sense that he or she complies with and does not question the current state of things. To be orderly in this sense is to be sufficiently docile to accept the world as it is and one's place within it. The historic and contemporary experience of working-class, racialized, and gendered populations demonstrates that police practices are designed to conform to and prioritise order, and that in the pursuit of order the exercise of police power is defined by violence. The violence that defined the original production of capitalist social relations is mirrored by the violence required to sustain them, and police violence needs to be understood as part of the normal exercise of police power in the interests of the state and capital.

Protest policing, as the regulation of political and social movements, is central to, rather than an aberration from, this general function of police. Collective resistance and political struggle have been the focus of police from its origins and to suggest that the violent suppression of struggles for economic, political, and social justice is evidence of 'exceptional' policing is to ignore history. Policing of protest is, in other words, a pacification process *par excellence*, one that demonstrates both the productive and destructive dimensions of the exercise of police power. As Berksoy illustrates in this volume, the police response to protest aims at both the protest/movement's centre *and* its peripheries. To put that another way: the immediate suppression of a particular disruptive event is only part of the wider process, and it is the wider process that is important, for it is that which aims to *produce* the ideal protester. This ideal protestor is one that is suitably disciplined, docile, and non-disruptive.

This process is illustrated in the police response to anti-fracking protests and laid bare in GMP's response to the Barton Moss camp. The police response to the camp aimed, through the use of arrests and restrictive bail conditions, to disrupt the protest, to limit its disruptive potential, and to ensure that IGas got its daily shipment of trucks. The police, referred to by many protectors as 'IGas's private army,' appeared to those involved, and to many observers, to be facilitating the drilling operation rather than the legal, peaceful protest. When we consider that at Barton Moss, much like at Balcombe, the majority of cases have subsequently resulted in charges being dropped, with cases discontinued or defendants found not guilty,[36] the policing operation appears to have been aimed at preventing the protest from being effective rather than responding to breaches of the law.

Yet such brazen police violence in the face of media attention (social media or otherwise) also sends a clear signal to those on the geographic or ideological peripheries of the opposition: that any protest against the operation of fracking is both illegitimate and dangerous. The overwhelming police presence, the regular, 'normalised' use of public order officers, and the well-publicised number of arrests, all suggest that such protest is inherently dangerous and in need of heavy policing. The careful PR management of the policing operation by GMP served to reinforce the idea that this style of policing was in direct response to the nature of the protest, reinforcing the well-rehearsed idea that violence is used by police in exceptional circumstances and only in response to violence. GMP consistently misrepresented the number of complaints against police officers[37] and where the force's spokespeople, including the Chief Constable, acknowledged the very physical style of policing, they sought to blame protesters for provoking and antagonising officers, thus reiterating the idea of a reactive policing operation and GMP's reluctant use of force.[38] In direct appeals to the public through press statements, Fahy sought to present an image of police officers under attack yet acting with restraint and patience in a "difficult situation." In his narrative, this was a situation created by protesters who were "personally insulting"[39] and "verbally abusive"[40] toward officers.

The majority of media sources regurgitated this police narrative, and underneath the repeated lines about balancing competing human rights claims, the attempt to delegitimise the protest was clear. In the police narrative, the camp residents and their supporters were consistently unreasonable in their demands of the council, the landowners, IGas, and the police. This account accorded with the then Prime Minister's own criticism of opponents to fracking[41] as well as the dominant representation of protesters who challenge the status quo as being outside of reason. The attempt to delegitimise the protesters' message as well as their (direct) actions was central to the police response and this type of PR strategy needs to be situated on a continuum of police violence. As Fernandez has argued, protest policing, characterised by the physical violence and intensive surveillance of protesters well documented in recent years, works "from the inside out, forcing compliance in the movement through fear rather than through direct hard-line repression."[42] It seeks to impose a vision of appropriate forms of political activism; direct action and a politics of the street are unacceptable. The acceptable response to injustice is instead reduced to signing an online petition and staying at home, putting one's faith squarely and solely in the parliamentary process; this construction shapes the behaviour of those within and beyond the protest.

There is, however, more to be said about the importance of policing to the (re)production of the current order, and this takes us to the heart of contemporary capitalist accumulation. To understand why, we need to consider the political economy of fracking.

FRACKING AS PRIMITIVE ACCUMULATION

Fracking is a form of primitive accumulation that demonstrates the enduring importance of this form of accumulation in the contemporary era, as well as its importance to the ongoing project of class warfare. If we avoid the trap of relegating so-called primitive accumulation to an 'original stage' of capitalism, it is possible to begin to see how the current bourgeois order is as reliant on the violence of enclosure and dispossession that Marx explained in the final chapters of *Capital*.[43] Focussing on the policing of opposition to fracking, and the pacification process inherent within, demonstrates only that police power remains as central to class warfare as it has always been.[44]

For many critical scholars considering contemporary forms of capital accumulation, the focus on processes of financialisation and their effects is understandable[45]; as the promise of regulation, given post-2008 by states and finance capital itself, is revealed to have been empty, the "speculative raiding carried out by hedge funds and other major institutions of finance capital"[46] continues apace. For Harvey writing in 2004, such speculative raiding was at the "cutting edge of accumulation by dispossession"[47] and the contemporary politics of austerity have certainly done nothing to reverse this. However, in the case of fracking we see another front on which accumulation is predicated on dispossession in a form more akin the original Enclosure Acts.[48] The basic form of fracking involves the physical enclosure of land and the privatisation of the natural commons in the interests of capital.

The physical act of privatising a space, of enclosing a site, is central to fracking. The necessary industrial infrastructure even for exploratory drilling requires the establishment of a site licenced to the corporation and, as opposition grows, the importance of barricading the drill site is heightened. We see here the importance of the fence, of the physical partition, in much the same way it was in the original formulation of the idea of private property.[49] The violence of enclosure, the raw act of the dispossession of land that has at least the appearance of being public,[50] remains central to the experience of those communities within which fracking takes place. Yet the acts of enclosure, exploitation, and degradation extend much farther than the fence would imply; the physical partition of the drill site is but the tip of the iceberg when fracking involves horizontal drilling technologies that have the capacity to extend up to 6000 feet from the well site.[51] This violent remaking of space has a visible and invisible element to it, and the potential for what Rob Nixon describes as the 'slow violence' of environmental damage[52] runs far beyond the partitioned site above ground. It is this new technological fusion of hydraulic fracturing and horizontal drilling that has enabled the exploitation of previously inaccessible or unprofitable shale reserves. This fusion is also the focal point of many campaigners who point to the reach of fracking companies in terms of the land they effectively colonise and the potential negative impact that they have. Debates about who is 'local' to fracking sites as a way to delegitimise protesters who have travelled to fracking sites—a line pushed by police at Barton Moss[53]—wilfully ignore the fact that the potential for air, land, and water pollution (as well as the contribution to climate change) means that we are all 'local'; the potential impacts know no bounds.

As a contemporary form of primitive accumulation, fracking relies upon state violence to impose these processes in the face of popular opposition. In considering contemporary forms of accumulation by dispossession, Harvey stressed that these processes require the power of the state to impose them upon communities, by force if necessary, and that this is a key indicator of the historical continuities of primitive accumulation.[54] The reliance upon police power to pacify protest is illustrative of this dependency of capital on state violence. What we see in this context is the ongoing war of accumulation waged by the state and corporations.[55] And yet these continuities are also evident in the fact that primitive accumulation provokes resistance. We must understand anti-fracking movements—and anti-globalisation movements as Harvey suggests are another key example[56]—as being driven by an opposition to the violence of primitive accumulation. Their opposition needs to be read here as resistance to the very foundation of capital, and they are responded to by states and corporations in line with the threat that they pose.[57] The policing of these movements is indicative of this threat, and in the case of anti-fracking campaigns, even those parts of the movement that position themselves as 'moderate,' or claim to be apolitical, are anti-capitalist to the extent that they oppose this process. To take a position on fracking implies that one has adopted a direct role in class warfare, and for many protesters new to direct activism at Barton Moss, this was the first time they had seen the function of police power at the heart of this war.

Moreover, fracking signifies the endurance of primitive accumulation as the historical foundation of capital at a further level if we consider how the interrelated ideas of waste and improvement frame the official position. Neocleous demonstrates how waste has been a bourgeois obsession from Locke onwards that can only be responded to through "improvement."[58] While idle labour has always a focal point of this obsession with waste—all too obvious in contemporary neoliberal discourses on welfare and 'worklessness'—uncultivated *land* has long troubled bourgeois thinkers for this same reason. If land is not worked and no value is extracted then it is, according to this bourgeois logic, wasted, and this failure must be corrected through a strategy of improvement. The logic of improvement was not only central to Locke's political philosophy but had by the eighteenth century become "a key category of capitalist modernity" that came to "underpin the whole enclosures movement."[59] Primitive accumulation as the colonisation of common land was driven by the idea of improvement. We want to suggest that it is possible to see this logic continuing to underpin the contemporary enclosures movement visible through fracking.

Lord Howell's now infamous statement that fracking in the UK should be directed toward the "uninhabited and desolate areas" of the north-east of England captures the contemporary appeals to waste and improvement.[60] For Locke, land that was "left wholly to Nature"[61] was waste and required improvement; for Howell, much like his former colleagues in the government, those areas of land in the northeast of England that do not have the real estate value of the southeast, but which do have shale reserves, are waste while they are left unexploited. Despite Howell's appeals to notions of environmental protection (and those by the current government), it seems that when it comes to fracking, the state and capital see a piece of Green Belt land in the north of England—as found at Barton Moss—in the way that Locke saw those tracts of uncultivated land in seventeenth century England and America. The notion of uncultivated land in the current context takes on an added impetus when the apparent riches below the surface in the form of fossil fuels are exposed. To leave this gas in the ground would be in Cameron's terms as 'irrational' as it would be to sustain waste and to oppose improvement that comes in the form of 'economic growth' and 'energy security.'

The argument for economic growth is bound up in the promise of jobs in fracking areas. The shale industry, backed by national government, have made assurances that those areas of the country that support fracking will see thousands of new jobs[62] and this has been translated into a direct appeal at the local level.[63] In a post-industrial city like Salford, which has suffered from some of the worst effects of austerity and one of the highest unemployment rates in the country,[64] this argument for fracking has great appeal. The economic imperative in the original enclosures—" 'setting free' the agricultural population as a proletariat for the needs of industry"[65]—continues into the present in the form of 'economic regeneration.' The promotion of fracking through a PR strategy jointly run by government and the shale gas industry aims to win hearts and minds and undermine the opposition.[66] Opponents to fracking are presented as the enemies of improvement. And as the enemies of improvement, they are the enemies of capital.

The idea of the "improvement of waste" was also historically "used to challenge customary rights,"[67] and fundamental property rights in the UK are being transformed through the contemporary appeals to improvement that underpin the arguments for fracking. The UK government has recently introduced major changes to trespass laws in the Infrastructure Act 2015 to facilitate fracking companies in their access to shale reserves located under private property. These changes were initially put out to consultation in May 2014 and 99% of responses were opposed.[68] The changes relate to the fact that in the UK companies previously needed to obtain a right of access from landowners for both surface and underground land. Landowners in the UK have historically had rights to land at the surface and, quite literally, down to the centre of the Earth; landowners within the previous legal framework could therefore object to the use of land below their property. In the case of horizontal drilling, the companies could well need to access land under the property of numerous consenting landowners to avoid committing trespass.

This left open the opportunity for a single landowner with rights to a small piece of land to object and compromise the viability of a fracking venture. As the government explained:

> The fundamental issue with the present system is that it gives a single landowner the power to significantly delay a development regardless of how others in their community feel about it (including other landowners), and even though the drilling and use of underground wells does not affect their enjoyment of their land.[69]

The changes to trespass laws in the Infrastructure Act 2015 revoke this right of objection by limiting rights to underground land to a depth of 300 meters and in doing so radically transform property rights. The consultation report made clear that the motivation for these changes was to challenge the disruptive potential of opponents to fracking who sought to utilise the law and exert property rights in their strategies for protest:

> Some landowners are likely to want to use this process to delay shale development in their local area, despite the fact that the community as a whole may be supportive. This method of delaying development has already been proposed by groups opposed to fracking: in 2013, they suggested the idea of buying up "ransom strips" in an attempt to use these to refuse to grant access rights underneath the drilling route.[70]

Long standing property rights that are open to this type of 'abuse'—holding capital to 'ransom'—cannot be allowed to stand.

These changes to law are central to the establishment of a regulatory framework in which the industry can be quickly established and gas extraction started. The UK government have sought to lobby regulators at the EU level and stop attempts to set legally binding environmental regulations for the shale gas industry,[71] with Cameron mobilising the well-established mantra that regulation is a barrier to investment. In addition, the government proposed changes to the system of business rates through which councils will be entitled to keep 100% of business rates raised from fracking sites.[72] This essentially operates as a bribe to those authorities who grant licences and who are now several years into an austerity program that has seen their access to central government funding plummet. We see here the state acting solely in the interests of capital utilising its power to revise and reduce regulation order to encourage investment and remove obstacles to accumulation.

The collaboration between the state and corporations runs yet further here in promotion of primitive accumulation. The UK fracking industry's representative body United Kingdom Onshore Operators Group (UKOOG) launched a 'Community Engagement Charter' in June 2013. Alongside further commitments to consultation and effective regulation, the charter details the 'benefits' from fracking that will be passed on to local communities. These include the provision of "benefits to local communities at the exploration/appraisal stage of £100,000 per well site where hydraulic fracturing takes place" and the commitment to "provide a share of proceeds at production stage of 1% of revenues, allocated approximately 2/3rd to the local community and 1/3rd at the county level."[73] These commitments have since been cited in government policy documents on planning permission for fracking[74] and lauded by ministers as being illustrative of the industry's desire to establish a "partnership with local people."[75] Yet the assumption that 1% is sufficient recompense for the enclosure of public space and privatisation of the natural commons illustrates the central place, and unquestioned status, of primitive accumulation in contemporary capitalism. The obligation to provide minimal recompense to local communities is to be combined with historically low tax rates for onshore shale gas production[76] as state and capital combine forces in the interests of accumulation.

What this arrangement also demonstrates is the obvious contempt of fracking corporations and the government for those who will experience the negative impacts of fracking; this includes both the wider community, who will suffer the various negative effects of the industrial process and its environmental impacts, and those property owners under whose homes the ground is to be fracked. The beneficiaries in this situation are obviously not the local community but state and capital as the class war continues. While some of those on the losing side here are propertied landowners, the basic features of the class war have not changed; primitive accumulation remains pivotal to the accumulation of capital, and those who mount an opposition, propertied or not, will not be tolerated. However, it is important to note that as a further historical continuity in the class war, it is those *without* property, those who mount their opposition from the street, who feel the full force of the violence of the state.

PACIFICATION AND STRUGGLE

The centrality of state power to primitive accumulation is as central now as it has ever been. Fracking demonstrates the permanence of the war of accumulation waged by state and corporations, and this is being fought on numerous fronts across the globe. We have sought here to provide a brief exposition of the two main dimensions of this as it unfolds through the UK fracking industry. In the first instance, forging a new regulatory regime to encourage the move by corporations from exploration to extraction is of great importance. Changing property laws, opposing EU regulation, launching a new bidding round for licences,[77] and providing government-funded boreholes[78] are all examples of state interventions located on a continuum of state violence. That these interventions come in the face of massive opposition to fracking is not surprising; the state has taken a lead role— as history would have us expect—in pacifying resistance. The response to opposition needs to be understood as part of the state's role in the war of accumulation; the exercise of police power in the drive to pacify resistance is of fundamental importance and the role of the state becomes increasingly important as opposition grows. Pacification is, as it has always been, vital to capital accumulation.

In recent years the policing of protest in nations across the globe has developed through the use of enhanced police powers introduced in the name of security.[79] The mobilisation of counter-terrorism powers to combat political activism is all too attractive to states across the globe who are threatened by resistance movements. Criminalising protest through the conflation of dissent, extremism, and terrorism has been effective in attempts to delegitimise and disarm political and social movements that seek to disturb the status quo.[80] Despite this, the policing at Barton Moss is not radically new: it does not signify a major departure from the type of policing historically experienced by social and political movements around the world, including in the UK. Nor does it suggest that protest policing itself has substantively changed in recent years, despite the repeated claims in the academic literature.[81] Barton Moss confirms what activists around the world continue to describe, that police violence remains a key tool of state responses to effective protest. We must however, be aware that this violence is at times selectively applied, and we must be aware that it has both destructive and productive dimensions.[82]

What the Barton Moss case study does demonstrate is a number of important dynamics to the response to anti-fracking protests that should inform ongoing work on pacification. In the first instance, the experience at Barton Moss demonstrates the role of the state in facilitating primitive accumulation, and that the exercise of state violence will keep pace with opposition as it grows. We need to consider the apparent departure from 'normal' policing in relation to the history of police responses to struggle. Barton Moss demonstrates that pacification *always* presumes some form of resistance. As the capitalist order is planned and enforced with primitive accumulation as ever at its foundation, it is also resisted. Pacification is the continuum of police violence on which this takes place and in considering both the repressive and productive dimensions to the exercise of police violence, we must consider the points at which it is resisted in myriad ways and with varying degrees of success.

We must therefore also consider the function of the representation of resistance by state and capital. The appeals to the public by the police—the (re)production of the police narrative that positioned the protesters as the problem—were aimed at winning over 'hearts and minds' as has always been central to the pacification process.[83] At Barton Moss we saw both the iron fist and velvet glove of police power; the PR campaign and physical repression illustrate the two sides of pacification, its productive and destructive dimensions.

The continuous reiteration of the acceptance of 'peaceful' protest by police, politicians, and media commentators, draws a line between acceptable and unacceptable forms of protest. We see regularly that acceptable forms of protest, those defined as 'peaceful,' include only those actions that do not threaten the status quo. Any real attempt to disrupt or even bring into question the fundamental features of the current social order fall outside the incredibly narrow definition of 'peaceful' protest and are thus defined as unacceptable, and responded to as such. As Starr et al. suggest, in light of the false dichotomy between "good" and "bad" protesters created by this framing, "all those who disturb in the slightest the channel provided by the police are threats, are violent, unpredictable, preternaturally out of control, beyond the bounds of social mores."[84] The emphasis on respecting the right to peaceful, lawful protest enables the police to justify the repression of protests that they can designate as outside of the accepted parameters: "the fetish for the 'peaceful protester' should be understood as a technique of pacification that conceals and fortifies the class violence of capitalism."[85] This is as evident in responses to anti-fracking protests in Salford as it has been in response to student protests, as well as protests against globalisation, climate change, war, austerity and the far right in recent years.

In confronting primitive accumulation through direct action (and highlighting the dangers involved in the extraction of unconventional fossil fuels), fracking protesters are stepping outside of the incredibly limited official understanding of legitimate 'peaceful' protest. In doing so their actions have the effect of disturbing the wider social order, in which capitalism, sustained through both accumulation by dispossession and a dependence on fossil fuels, is sealed off from any real alternatives. Anti-fracking activists have questioned the immediate use of specific sites for drilling as well as the broader energy policies that sustain a carbon intensive capitalism; the 'greening' of capitalism, a key legitimating tool utilised by states and corporations, is here exposed as a charade. In addition, they have highlighted and opposed the collusion between state and corporations in relation to fracking policy in the UK that has been exposed at the national level[86] as well as in the local context at Barton Moss.[87]

The protest camp itself is a clear sign of 'disorder,' symbolising an opposition to state-corporate collusion in the exploitation of the natural environment. The very nature of activists, experienced and inexperienced, choosing to live in a makeshift camp by the side of a minor road, disturbs the usual state of things in the local context but also demonstrates the seriousness of the issue at hand. It is this disorderly nature of the camp at Barton Moss—not to be confused with a suggestion that the camp was run in a disorderly fashion—that gives rise to the policing response. The very public nature of protest and the immediate effects of direct action explicitly challenge the parliamentary process' monopoly on legitimacy. Protesters both demand and *enact* a form of politics that refuses the compromise inherent in parliamentary democracy. As Rimke suggests, we need to understand the policing of protest as precisely a response to this challenge:

> Policing of public protest reifies the highly controlled and restrictive processes of parliamentary democracy as the preferred form of political engagement and expression. Alternative forms of politics that challenge the pacifying politics of parliamentarianism are thus represented through the dominant security doxa as irrelevant, absurd, pathological or even criminal.[88]

In relation to fracking, the activists' approach challenges the very foundation of the official presentation of the natural environment and its relationship to the current order; in doing so they directly challenge the construction of fossil fuels left in the ground as waste. The demands of many anti-fracking activists that so trouble Cameron and others are 'irrational' because they call for fossil fuels to be left in the ground; the demand is not for safer fracking or a more equitable distribution of the proceeds, but for it to be abandoned.

In their dominant framing, climate change and environmental issues more generally are comprehensively depoliticised. Climate change is defined in current policy primarily as a security issue and, as a result, it is closed to substantive political debate and any intervention that seeks to consider it in relation to the current political and economic order.[89] As Neocleous notes, "'securitising' an issue does not mean dealing with it politically, but bracketing it out and handing it to the state."[90] Anti-fracking activists are, then, in many cases, consciously seeking to reclaim the issue from the state and to redefine its causes and solutions. By doing so, they impose a (re)politicisation of the issue that challenges the state definition and in fact places the state and its corporate allies at the heart of the problem. In refusing the logic of security as it frames environmental issues, these activists are by definition a sign of disorder, and their refusal of political docility effectively forces the state to (re)turn to use of overt police power. Given the track record of police responses to populations designated as disorderly, and the importance of primitive accumulation to capitalist order, the police operation at Barton Moss is in line with what the history of policing (including its recent developments) should have us expect.

NOTES

1. David Berkowitz, "Natural Gas: Sale of the Century," *The Economist*, June 2, 2012, accessed November 25, 2014, http://www.economist.com/node/21556242.
2. Berkowitz, "Natural Gas: Shale of the Century."
3. Unconventional deposits are those distributed throughout a geologic formation over a wide area, but not in a discrete, easily accessible reservoir. See Gene Whitney, Carl E. Behrens, and Carol Glover, *US Fossil Fuel Resources: Terminology, Reporting, and Summary*, Congressional Research Service, November 30, 2012, accessed January 6, 2015, http://www.epw.senate.gov/public/index.cfm?FuseAction=Files.view&FileStore_id=04212e22-c1b3-41f2-b0ba-0da5eaead952.
4. Anna Willow and Sara Wylie, "Politics, Ecology, and the New Anthropology of Energy: Exploring the Emerging Frontiers of Hydraulic Fracking," *Journal of Political Ecology* 21 (2014): 222-236, accessed July 24, 2015.
5. I.J. Andrews, *The Carboniferous Bowland Shale Gas Study: Geology and Resource Estimation*, British Geological Survey for Department of Energy and Climate Change (London: Department for Energy and Climate Change, 2013).
6. Bernie Vining and Steve Pickering, *Petroleum Geology: From Mature Basins to New Frontiers – Proceedings of the 7th Petroleum Geology Conference* (London: The Geological Society, 2010).
7. Willow and Wylie, *Politics, Ecology, and the New Anthropology of Energy*.
8. Willow and Wylie, *Politics, Ecology, and the New Anthropology of Energy*, 223; See also Timothé Feodoroff, Jennifer Franco, and Ana Maria Rey Martinez, *Old Story, New Threat: Fracking and the Global Land Grab* (Transnational Institute: Amsterdam, 2013).
9. Robert W. Howarth, Anthony Ingraffea, and Terry Engelder, "Natural Gas: Should fracking stop?" *Nature* 477 (2011), 271–275.
10. BBC News, "Fracking Tests Near Blackpool 'Likely Cause' of Tremors," *BBC News*, November 2, 2011, accessed November 28, 2014, http://www.bbc.co.uk/news/uk-england-lancashire-15550458.
11. Michael Gross, "Dash for Gas Leaves Earth to Fry," *Current Biology*, 23, no. 20 (2013): R901-R904.

12 Willow and Wylie, *Politics, Ecology, and the New Anthropology of Energy*, 223.
13 David Harvey, "The 'New' Imperialism: Accumulation by Dispossession," *Socialist Register* 40 (2004): 63-87, accessed July 12, 2014, http://socialistregister.com/index.php/srv/article/view/5811#.VbIACv9REdU.
14 Feodoroff et al., *Old Story, New Threat*, 2.
15 Feodoroff et al. identify citizens' campaigns against fracking established by 2013 in the following countries: United States, Canada, New Zealand, Australia, Morocco, Algeria, Tunisia, South Africa, Spain, UK, Romania, Poland, Brazil, Chile, Bolivia, Mexico, India, Argentina, Sweden and Austria. Feodoroff et al, *Old Story, New Threat,* 12.
16 Mark Neocleous, "War as Peace, Peace as Pacification," *Radical Philosophy* 159 (2010): 8-17; Mark Neocleous, "A Brighter and Nicer New Life: Security as Pacification," *Social Legal Studies* 20 (2011): 191-208, accessed July 23, 2105; Mark Neocleous, "The Dream of Pacification: Accumulation, Class War, and the Hunt," *Socialist Studies/Études Socialistes* 9, no. 2 (2013): 7-31; Mark Neocleous, *War Power, Police Power,* (Edinburgh: Edinburgh University Press, 2014); Mark Neocleous and George Rigakos, *Anti-Security* (Ottawa: Red Quill Press, 2011); George Rigakos, "To Extend the Scope of Productive Labour: Pacification as a Police Project," in *Anti-Security*, eds. Mark Neocleous and George Rigakos (Ottawa: Red Quill Press, 2011); George Rigakos and Aysegul Ergul, "The Pacification of the American Working Class: A Time Series Analysis," *Socialist Studies/Études Socialistes*, 9, no. 2 (2013): 167-198.
17 The authors visited the site of the camp on a number of occasions and conducted a series of interviews with camp residents and visitors. This chapter does not make direct use of those interviews but draws upon the research conducted at Barton Moss since late 2013.
18 Neocleous, *The Dream of Pacification: Accumulation, Class War, and the Hunt*, 18.
19 Neocleous, *War Power, Police Power;* Neocleous, *The Dream of Pacification.*
20 Karl Marx, *Capital: A Critique of Political Economy*, Vol. 1, trans. B. Fowkes (Harmondsworth: Penguin, 1976); Neocleous, *War Power, Police Power;* Harvey, *The 'New' Imperialism.*
21 Red Pepper, "Balcombe Latest: 50 Days on," *Red Pepper blog*, September 13, 2013, accessed November 25, 2014, http://www.redpepper.org.uk/balcombe-latest-50-days-on/.
22 Danny Chivers, "The Frack Files," *New Internationalist* 468 (2013), 12-24.
23 For full explanation of lock-ons see Chapter 12 of Road Alert!, *Road Raging: Top Tips for Wrecking Roadbuilding,* (Newbury: Road Alert!, 1997), accessed January 6, 2014, http://www.eco-action.org/rr/index.html.
24 Joanna Gilmore, Will Jackson and Helen Monk, *Keep Moving! Report on the Policing of the Barton Moss Community Protection Camp*, Liverpool: CCSE, http://statewatch.org/news/2016/mar/uk-policing-barton-moss-fracking-protest-report-2-2016.pdf
25 Salford Star, "Solicitor Accuses Greater Manchester Police of Political Policing at Barton Moss," *Salford Star*, February 6, 2014, accessed November 16, 2014, http://www.salfordstar.com/article.asp?id=2131.
26 Greater Manchester Police, "Chief Constable's Statement on Barton Moss Protest," [Press Release], February 7, 2014, accessed September 20, 2014, http://www.gmp.police.uk/content/WebsitePages/6662DEE03DF784C680257C78003CF8A6?OpenDocument.
27 Greater Manchester Police, *Chief Constable's Statement on Barton Moss Protest.*
28 Peter Fahy, "Changes to Policing," *Chief Constable Sir Peter Fahy*, February 27, 2014, accessed July 25, 2015, http://gmpolice.wordpress.com/2014/02/, emphasis added.
29 Dan Thompson, "Solicitor Compiles Dossier for UN on police 'aggression' at Barton Moss fracking protests," *Manchester Evening News*, May 25, 2014, accessed November 30, 2014, http://www.manchestereveningnews.co.uk/news/greater-manchester-news/fracking-salford-solicitor-compiles-dossier-6874073.
30 Neocleous, *War as Peace, Peace as Pacification*; Neocleous, *A Brighter and Nicer New Life*; Neocleous, *The Dream of Pacification*; Rigakos, *To Extend the Scope of Productive Labour.*
31 Rigakos, *To Extend the Scope of Productive Labour,* 78.
32 Neocleous, *The Dream of Pacification.*
33 Neocleous, *A Brighter and Nicer New Life,* 198.
34 Neocleous, *War Power, Police Power*, 32.
35 Rigakos and Ergul, *The Pacification of the American Working Class*, 169.

36 Sandra Laville, "Sussex Police Under Fire for 'Criminalising' Fracking Protests," *The Guardian*, May 15, 2014, accessed August 8, 2014, http://www.theguardian.com/environment/2014/may/15/sussex-police-criminalising-fracking-protest-acquittals-balcombe; Salford Star, "29 Barton Moss Anti Fracking Cases Discontinued as Solicitor Slams GMP," *Salford Star*, June 18, 2014, accessed August 26, 2014, http://www.salfordstar.com/article.asp?id=2305.
37 A fact confirmed through Freedom of Information requests following the end of the camp. See Netpol, "Police and Fracking Companies – In Each Other's Pockets?" *The Network for Police Monitoring*, August 5, 2014, accessed September 25, 2014, https://netpol.org/2014/08/05/police-fracking-collusion/.
38 Fahy in Greater Manchester Police, *Chief Constable's Statement on Barton Moss Protest*.
39 Fahy in Greater Manchester Police, *Chief Constable's Statement on Barton Moss Protest*.
40 Kate Storey, "GMP Chief Sir Peter Fahy Slams Fracking Protesters for Trying to 'Provoke' Police Officers," *Manchester Evening News*, February 7, 2014, accessed January 6, 2015, http://www.manchestereveningnews.co.uk/incoming/barton-moss-fracking-greater-manchester-6683511.
41 Prime Minister David Cameron has described opponents to fracking as 'irrational' suggesting many are 'religiously opposed' to carbon based fuels. See Patrick Wintour, "Fracking opponents are being irrational, says David Cameron," *The Guardian*, January 14, 2014, accessed November 25, 2014, http://www.theguardian.com/politics/2014/jan/14/fracking-opponents-irrational-says-david-cameron.
42 See Luis Fernandez, *Policing Dissent: Social Control and the Anti-globalization Movement* (New Brunswick: Rutgers University Press, 2008), 77.
43 Marx, *Capital*; See also, Harvey, *The 'New' Imperialism*; Neocleous, *War Power, Police Power*, Chapter 2.
44 Mark Neocleous, *The Fabrication of Social Order* (London: Pluto Press, 2000); Neocleous, *War Power, Police Power*.
45 See Saskia Sassen, "A Savage Sorting of Winners and Losers: Contemporary Versions of Primitive Accumulation," *Globalizations*, 7, no. 1-2 (2010): 23-50, accessed July 23, 2015.
46 Harvey, *The 'New' Imperialism*, 75.
47 Harvey, *The 'New' Imperialism*, 75.
48 It may be more appropriate to see this as a related front given the role of finance capital in fuelling the fracking boom (Feoderoff et al., *Old Story, New Threat*), but the physical act of enclosure marks an important distinction at least in the direct experience, and awareness, of communities affected.
49 Neocleous, *War Power, Police Power*, 68-69.
50 The green belt land authorised for exploratory drilling at Barton Moss is privately owned by one of the North West's largest land owners, Peel Holdings, but for many involved in the opposition, the site was described as though the land were public and many emphasised its importance to the local community.
51 American Petroleum Institute, *Hydraulic Fracturing: Unlocking America's Natural Gas Resources* (Washington DC: API, July 2014).
52 Rob Nixon, *Slow Violence and the Environmentalism of the Poor* (Cambridge, MA: Harvard University Press, 2011).
53 Sarah Womack, "Three Out of Four Anti-fracking Protesters Are Not Local and Are Just There to 'Disrupt and Intimidate' Communities and the Police, Top Officer Says," *Daily Mail*, January 23, 2014, accessed November 28, 2014, http://www.dailymail.co.uk/news/article-2544758/Three-four-anti-fracking-protesters-not-local-just-disrupt-intimidate-communities-police-officer-says.html.
54 Harvey, *The 'New' Imperialism*.
55 Neocleous, *War Power, Police Power*.
56 Harvey, *The 'New' Imperialism*, 75.
57 On the policing of anti-globalisation protests see Amory Starr, Luiz Fernandez, and Christian Scholl, *Shutting Down the Streets: Political Violence and Social Control* (New York: New York University Press, 2011). On the role of police and corporations in policing protest movements see Eveline Lubbers, *Secret Manoeuvres in the Dark: Corporate and Police Spying on Activists* (London: Pluto, 2012).
58 Neocleous, *War Power, Police Power*, Chapter 2.
59 Neocleous *War Power, Police Power*, 65.
60 BBC News, "Fracking OK for 'desolate' North East, says Tory peer," *BBC News*, July 30, 2013, accessed

November 25, 2014, http://www.bbc.co.uk/news/uk-politics-23505723.
61 Neocleous, *War Power, Police Power*, 62.
62 Terry Macalister and Patrick Wintour, "Fracking Could Generate £33bn and 64,000 Jobs for UK," *The Guardian*, April 24, 2014, accessed January 6, 2015, http://www.theguardian.com/environment/2014/apr/24/fracking-generate-investment-jobs-industry-report-uk.
63 Amy Glendinning, "3,500 jobs and £10bn Boost if North West Embraces Fracking," *Manchester Evening News*, July 21, 2014, accessed January 7, 2015, http://www.manchestereveningnews.co.uk/news/greater-manchester-news/fracking-3500-jobs-10bn-boost-7462725.
64 Salford Star, "Salford Job Horror Gets Worse," *Salford Star*, August 27, 2013, accessed January 7, 2015, http://www.salfordstar.com/article.asp?id=1922.
65 Marx, *Capital*, 886.
66 Emails between shale gas executives and government officials released through Freedom of Information requests illustrate that the claims around numbers of jobs that will be created rely on industry figures that have then been repeated by government despite the figures being more than double the forecasts made by the Department of Energy & Climate Change. See Damian Carrington, "Emails Reveal UK Helped Shale Gas Industry Manage Fracking Opposition," *The Guardian*, January 17, 2014, accessed November 28, 2014, http://www.theguardian.com/environment/2014/jan/17/emails-uk-shale-gas-fracking-opposition.
67 Neocleous, *War Power, Police Power*, 65.
68 Damian Carrington, "Fracking Trespass Law Changes Move Forward Despite Huge Public Opposition," *The Guardian*, September 26, 2014, accessed November 20, 2014, http://www.theguardian.com/environment/2014/sep/26/fracking-trespass-law-changes-move-forward-despite-huge-public-opposition.
69 Department of Energy and Climate Change, *Underground Drilling Access* (London, DECC, May 2014), 21.
70 Department of Energy and Climate Change, *Underground Drilling Access*, 21.
71 Damian Carrington, "UK Defeats European Bid for Fracking Regulations," *The Guardian*, January 14, 2014, accessed November 23, 2014, http://www.theguardian.com/environment/2014/jan/14/uk-defeats-european-bid-fracking-regulations.
72 Nicholas Watt, "Fracking in the UK: 'We're Going All Out for Shale,' Admits Cameron," *The Guardian*, January 13, 2014, accessed November 20, 2014, http://www.theguardian.com/environment/2014/jan/13/shale-gas-fracking-cameron-all-out.
73 United Kingdom Onshore Operators Group, *Community Engagement Charter: Oil and Gas from Unconventional Reservoirs*, June 2013, accessed March 3, 2015, http://www.ukoog.org.uk/images/ukoog/pdfs/communityengagementcharterversion6.pdf.
74 Department of Energy and Climate Change, *Fracking UK Shale: Planning Permission and Communities* (London: DECC, February 2014).
75 Department of Energy and Climate Change, *Estimates of Shale Gas Resource in North of England Published, Alongside a Package of Community Benefits*, [Press Release], June 27, 2013, accessed March 3, 2015, https://www.gov.uk/government/news/estimates-of-shale-gas-resource-in-north-of-england-published-alongside-a-package-of-community-benefits.
76 The UK government have set the tax rate for shale gas extraction at 30%–less than half the 62% levied on North Sea oil production–to facilitate the process and of course to increase profit. The potential revenues per well site over the 20-year life span are estimated to be between £240m and £480m by independent assessors, and between £500m and £1bn by government. See AMEC Environment & Infrastructure UK Limited, *Strategic Environmental Assessment for Further Onshore Oil and Gas Licensing: Environmental Report* (London: DECC, December 2013); Emily Gosden, "Government Accused of 'Overhyping' Shale Gas Benefits," *The Telegraph*, January 17, 2014, accessed March 3, 2015, http://www.telegraph.co.uk/news/earth/energy/10580255/Government-accused-of-overhyping-shale-gas-benefits.html.
77 BBC News, "Fracking in Beauty Spots in Exceptional Circumstances," *BBC News*, July 28, 2014, accessed November 20, 2014, http://www.bbc.co.uk/news/uk-28520323.
78 Damian Carrington, "Taxpayers to fund hundreds of fracking boreholes across the country," *The Guardian*, November 22, 2014, http://www.theguardian.com/environment/2014/nov/22/taxpayers-fund-hundreds-fracking-boreholes.
79 Lesley, J. Wood, *Crisis and Control: The Militarisation of Protest Policing* (London: Pluto Press, 2014); Joanna Gilmore, "Policing Protest: An Authoritarian Consensus," *Criminal Justice Matters* 82, no. 1 (2010): 21-23; Fernandez, *Policing Dissent*; Starr et al., *Shutting Down the Streets*.

80 Will Jackson, "Securitization *as* Depoliticisation*:* Depoliticisation as Pacification," *Socialist Studies/Études Socialistes* 9, no. 2 (2013): 146-166; Jeffery Monaghan and Kevin Walby, "Making Up 'Terror Identities': Security Intelligence, Canada's Integrated Threat Assessment Centre, and Social Movement Suppression," *Policing and Society,* 22, no. 2 (2012): 133-151, accessed July 20, 2015; Colin Salter, "Activism as Terrorism: The Green Scare, Radical Environmentalism and Governmentality," *Anarchist Developments in Cultural Studies* 1 (2011): 211-238, accessed July 24, 2015, http://www.anarchist-developments.org/index.php/adcs_journal/article/view/36/37.

81 Recent academic literature has suggested that protest policing has been transformed to a new consensus based model in which the police violence all too familiar in previous decades has been replaced by negotiation and facilitation. For recent literature on this view in the UK, see Hugo Gorringe and Michael Rosie, "'We *Will* Facilitate Your Protest': Experiments with Liaison Policing," *Policing* 7, no. 2 (2013): 204-211; Hugo Gorringe, Clifford Stott, and Michael Rosie, "Dialogue Police, Decision Making, and the Management of Public Order During Protest Crowd Events," *Journal of Investigative Psychology and Offender Profiling* 9, no. 2 (2012): 111-125; David Waddington, "A 'Kinder Blue': Analysing the Police Management of the Sheffield Anti-'Lib Dem' Protest of March 2011," *Policing and Society: An International Journal of Research and Policy* 23, no. 1 (2013): 46-64. See also Donatella della Porta and Herbert Reiter, *Policing Protest: The Control of Mass Demonstrations in Western Democracies* (Minneapolis: University of Minnesota Press, 1998); Donatella della Porta, Abby Peterson, and Herbert Reiter, *The Policing of Transnational Protest* (Farnham: Ashgate, 2006).

82 See Fernandez, *Policing Dissent*; Starr et al,, *Shutting Down the Streets;* Jackson, *Securitization* as *Depoliticisation.*

83 Neocleous, *A Brighter and Nicer New Life,* 197; Richard A. Hunt, *Pacification: The American Struggle for Vietnam's Hearts and Minds* (Boulder, CO: Westview Press, 1995); See also Toews, this volume; Wall, this volume.

84 Starr et al,, *Shutting Down the Streets,* 95.

85 Heidi Rimke, "Security: Resistance," in *Anti-Security*, eds. Mark Neocleous and George Rigakos (Ottawa: Red Quill Press, 2011), 206.

86 Carrington, *Emails Reveal UK Helped Shale Gas Industry Manage Fracking Opposition.*

87 Netpol, *Police and Fracking Companies–in Each Other's pockets?*

88 Rimke, *Security: Resistance,* 209.

89 Jackson, *Securitization* as *Depoliticisation.*

90 Mark Neocleous, *Critique of Security* (Montreal and Kingston: McGill-Queen's University Press, 2008), 186.

Part 2
SPACE AND PACIFICATION

Humanizing Domination
On the Role of Urbanism in French Colonial Pacification Strategies[1]

Parastou Saberi

Urbanism must be implanted first in the hearts of men.

Marshal Hubert Lyautey, Resident-General of Morocco, 1912-1925

And thus, little by little, we conquer the hearts of the natives and win their affection, as is our duty as colonizers.

Joseph Marrast, French architect upon visiting Casablanca in 1920

The re-emergence of counterinsurgency in the mid-2000s shifted the attention of the imperial states to the strategic utility of civilian forms of interventionism for sustaining imperial domination and occupation.[2] Counterinsurgency, or what historically has been known as pacification, has since been advocated as the most effective way to install liberal stabilization and governance locally and globally. Despite its 'humanitarian' and 'peace-building' appearance, pacification is essentially a counter-revolutionary strategy, through which war is codified as peace. Critical analysis of contemporary counterinsurgency has explored the ways military strategists build upon the 'productive' dimension of war by systematically integrating security politics, development, and humanitarianism. The political goal is to nullify movements of resistance and to produce liberal subjects, thus demobilizing the collective will of the (occupied) population to oppose domination.[3]

In their attempt to combine liberal humanism and *matchpolitik*, contemporary North American and West European military strategists have been mining lessons from the late colonial pacification strategies.[4] This recent revival of colonial pacification in our conjuncture requires a deeper examination of the many dimensions of colonial pacification strategies. In this chapter, my goal is to shift attention to the spatial dimension of pacification by highlighting the imperative role of urbanism in 'humanizing' colonial pacification and domination. Scholars in urban geography and architectural history have already demonstrated the crucial role of colonial urbanism in materializing colonial power and transplanting colonial-capitalist socio-spatial relations.[5] Nonetheless, the link between pacification and colonial urbanism has remained unexamined. While the strategic use of town planning in colonial pacification goes back to the sixteenth century, the systematic appropriation of urbanism as a strategy of pacification with the aim of moderating the violence of colonial domination was born out of the social reform revolution of the nineteenth and early twentieth centuries, particularly in the fields of architecture (and later urban planning) as well as colonial warfare.

My engagement with the relationship between urbanism and pacification is both theoretical and historical. Building upon the recent work of Mark Neocleous, I suggest that his attempt to historicize pacification in relation to capitalist social totality needs to be pushed further by a closer examination of the role of state-bound spatial and territorial strategies and the racial dimension of these strategies in the ongoing processes of domination and accumulation. I do so by bringing into conversation the works of Neocleous, Henri Lefebvre and Frantz Fanon. This theoretical framework, I suggest, is useful for highlighting how territorial relations of domination and accumulation and the racial dimensions of these relations play vital roles in shaping the specific ways that colonization, pacification, urbanism and capitalism operate dialectically.

In the last section, I mobilize a Lefebvrian-Fanonian lens to look at the relationship between colonization, colonial urbanism and pacification in the context of the French colonies during the late nineteenth and early twentieth centuries. My specific focus here is on the pacification strategies of Louis-Hubert Lyautey, the first French Resident-General of Morocco from 1912 to 1925. Lyautey's pacification legacy is important because it shapes one of the significant historical moments in the collusion of military and spatial strategies to the point of forming some of the main pillars of colonial urbanism in the twentieth century. Lyautey indeed figures large (albeit separately) in two sets of literature: 1) the literature on colonial warfare[6] and 2) the literature on colonial urbanism.[7] By bringing together these two sets of disciplinary literature I show how a form of liberal "humanist urbanism"[8] shaped the core of Lyautey's pacification strategy that aimed to sustain colonial domination by codifying domination as the peaceful tolerance of the colonized "culture" and "way of life."

Lyautey's legacy has political and analytical implications for our understanding of the contemporary revival of pacification. His legacy helps us understand that in analyzing pacification we need to take into account the relational formations of state urban strategies, ideology, and (neo-) colonial spatial relations of domination and accumulation. It is through such relational formations that the state and state-like knowledge and symbolism[9] become strong forces in producing pacified spaces and subjectivities. This is imperative both in terms of historicizing strategies of colonial pacification and understanding our current conjuncture. The contemporary shift toward a more 'humanitarian' way of war and development is not limited to the zones of neo-colonial occupations and extractions. Parallel ideological shifts have been influential in re-defining the strategies of state-bound intervention in the urban 'priority zones' at the heart of imperial metropoles, where the majority of the non-White working class population reside. To conceive urbanism as a potential strategy for pacification that aims to re-order the relations of force can deepen our understanding of how in our 'post-colonial' times of neo-colonial wars state-bound strategies of urban intervention and policing function as a modality of (neo)-colonial pacification both at the heart of the imperial metropole and in the former colonies.

WAR, PACIFICATION, AND THE SURVIVAL OF CAPITALISM

The critical analyses of contemporary counterinsurgency strategies are mainly concentrated in the fields of critical security, IR, and international development studies. Inspired by Foucauldian theoretical perspectives, scholars have examined the rationalities behind the population-centric techniques of counterinsurgency.[10] While the political nexus of security-development has a long history going back to the birth of capitalism,[11] it has been argued that by deepening this historical link, the re-emergence of counterinsurgency has brought a strategic shift towards an increasing *"civilianization* of warfare,"[12] a shift that is based on the inextricably intertwined ways of liberal rule and war.[13] These analyses have been illuminating for highlighting the ways military doctrines mobilize liberal notions of security, peace, development, humanitarianism, governance, and stability to sustain imperial domination over "ungoverned spaces." Yet, in their sole emphasis on population and the disseminative and biopolitical character of power, these studies have left unexamined the role of the geopolitical and spatial relations, as well as the structural dynamics of imperialist capitalism.[14]

In contrast, radical geographers have highlighted the imperative of the socio-spatial and territorial relations of imperialism as a systemic feature inherently bound up with the development of capitalism.[15] Imperialist capitalism, as David Harvey reminds us, arises out of "invoking a double dialectic of, first, the territorial and capitalist logics of power and, second, the inner and outer relations of the capitalist state."[16] As Marxist analyses of imperialism remind us, the fundamental objective of the US-led 'long war' and its global pacification campaign is to globalize and secure the capitalist mode of production.[17] Seen from this perspective, the 'long war' is a continuation of what Marx and Engels called the "social war" – the "civil war" essential for securing bourgeois civil society politically, subjugating labour, land and everyday life to capitalist social relations.[18]

While the forms of the dominant conquest strategies of imperialism have changed depending on the historical and geographical specificities of conquests, what has remained unchanged, despite changing in form, is the territorial-symbolic project of mapping that resides at the heart of the (geo)political world visions dominating politics at various scales, from the international to the local. At the heart of the logic of targeting the so-called "ungoverned spaces" in need of (paternalist and imperialist) intervention reside Darwinist conceptions of development, understood as an evolutionary adaptation to capitalist relations and integration into imperialist capitalist order.[19] From Hugo Grotuis to John Stuart Mill, colonial theoreticians mobilized a colonial narrative of 'civilized/barbaric/savage' to justify war and colonization in the name of peace and progress, to secure capitalist accumulation.[20] In many ways, it is the continuation of this ideological narrative that one finds in the influential contemporary geopolitical narratives advocated by the likes as Robert Kaplan, Mary Kaldor and Thomas Barnett, for whom it is the endemic re-imagining of underdevelopment as danger(ous) that stands as an obstacle to maintaining the imperial peace and the will of the market.[21] We are faced with a world divided into a "core" (composed of an integrated metropolitan system of "stable" states) and "gaps" (composed of peripheral geographies marked as 'ungoverned spaces' of "failed" and "failing" states, characterized by underdevelopment, violence, crime, and corruption).

Importantly, such territorial-symbolic depictions of "ungoverned spaces" that animate politics of the 'long war' have analogous ideological forms that have been influential in shaping the politics of "social war" in the Global North.[22] In the imperial metropoles, what Loïc Wacquant[23] has called "territorial stigmatization" has become simultaneously fundamental to marginalizing poor neighbourhoods populated mostly by non-White working class population from former colonies, and to justifying state intervention in these spaces. These spaces are conceived by the state and the public (including armies of professionals) as bastions of violence and social dissolution, and they are consequently perceived as threats to the peace, order, and security of bourgeois urban spaces. Think of the *banlieues* in France, the *quatrieri degradati* in Italy, the "black ghettos" in the United States, the "sink estates" in Great Britain, the "tower neighbourhoods" in Toronto (Canada), and the *favelas* in Brazil. These so-called ungoverned urban spaces are increasingly subjected to intertwined state-bound strategies of re-development and policing. Whether at the urban or the worldwide scales, the conceptions of "ungoverned spaces" rely on two distinct yet complementary representations of space: as humanitarian spaces in need of tutelage and aid, and as danger zones in need of being secured.

Given these ideological parallels in shaping the politics of the "social war" in the imperial metropoles and that of the 'long war' worldwide, the re-emergence of pacification doctrine, I suggest, also needs to be situated in relation to re-ordering "the relations of force" at various moments and scales. These relations are also deeply spatial, as Antonio Gramsci notes.[24] Seen from this perspective, two interrelated questions deserve critical attention: what is the relationship between pacification and the survival of imperialist capitalism? And what is the role of space and spatial strategies in the materialization and reification of pacification? These are important questions, not least because they lead us to a sharper political excavation of how the material grounds for historical blocs form and re-form, and how the ideological dimension of their hegemonic claims are shaped and mobilized. In this context, the concept of pacification is crucial.

By tracing the theory and practice of pacification back to the colonial wars of conquest, Neocleous situates pacification in relation to the processes of colonization, the transformation of the state and the structural violence through which capitalist order has been constituted and accumulation secured. Understanding primitive accumulation as an ongoing process and the violence of accumulation as a form of war allows us to bring together the state's war power and police power in the processes of pacification, hence making connections between the "internal pacification" in the metropole and the pacification of colonies.[25] Neocleous conceptualizes pacification as a "productive" force in constructing a new social order by re-organizing everyday life and the social relations of power around a particular regime of accumulation, while, crushing opposition to that construction.[26] Pacification, Neocleous suggests, is a police project:[27] a form of political administration targeting "problem" populations, integral to the ways through which the state constitute and "secure" civil society politically.

Marxist analyses of imperialist war and Neocleous's contribution direct us to how war *and* pacification are integral to the survival of imperialist capitalism. Critical engagement with pacification thus needs to be situated in relation to the workings of hegemony by bringing into analysis not just the relations of accumulation, but also the relation of domination. The concept of pacification, however, needs to be spatialized. Capitalism, colonization, war, and policing are all spatial processes, as we well know.[28] As I will argue, re-organizing the territorial and spatial relations of domination has always been important to materializing, reifying, and sustaining pacification.

COLONIZATION, PACIFICATION, AND SPACE

The birth of pacification as a strategy of warfare and conquest goes back at least to the late sixteenth century.[29] In 1573, Phillip II of Spain replaced the idea of conquest with the idea of pacification.[30] This early, yet systematic, "humanization" of colonization took place at a conjuncture in which liberalism found in (international) law a way to capitalize on the "productive" dimension of war by codifying war as peace and security.[31] In the same year (1573), Phillip II also put forward the "Prescriptions for the Foundations of Hispanic Colonial Towns," which eventually gave birth to the grid pattern in town (and later urban) planning.[32] In *The Production of Space*, Henri Lefebvre captures this moment in the production of the colonial town, which is worth quoting at length:

> The Spanish-American town was typically built according to … the veritable code of urban space constituted by the *Orders of Discovery and Settlement*, a collection, published in 1573. … These instructions were arranged under three heads of discovery, settlement, and pacification. *The very building of the towns thus embodied a plan which would determine the mode of occupation of the territory and define who it was to be recognized under the administrative and political authority of urban power.* … The result is a strictly hierarchical organization of space, … a high degree of segregation is superimposed upon a homogenous space. … this artificial product is also an instrument of production: a *superstructure foreign to the original space serves as a political means of introducing a social and economic structure in such a way that it may gain a foothold and indeed establish its 'base' in a particular locality.* … The relation between the 'micro' (architectural) plan and the 'macro' (spatial-strategic) one does exist there … The main point to be noted, therefore, is the production *of a social space by political power – that is, by violence in the service of economic goals.*[33]

Lefebvre's observation points to the imperative of spatial dimensions in what Neocleous depicts as the birth of pacification. Phillip II's prescriptions became the predecessor for the integration of urbanism and urbicide into the politics of (colonial) occupation and pacification for centuries to come. The literature on colonial urbanism has firmly illuminated the centrality of urbanism in transplanting the capitalist mode of production in the colonies[34] and the relational formation of colonial and metropolitan spaces by focusing on the spatialization of (colonial) knowledge and power.[35] Despite shortcomings,[36] the upshot of this literature is an empirical affirmation of what Frantz Fanon[37] reminded us more than half a century ago. That is, the process of colonization is a deeply spatial process.

One of Lefebvre's main contributions to Marxist theory is to de-fetishize space. Focusing on the active processes of the production of space, he shows that spatial form is not only a powerful social force but also a product of temporal processes, strategies, and projects.[38] Understanding the state as an institutional condensation of social power, he emphasizes the presence of the state (state-like thinking and symbolism) in everyday life and underlines how the state plays a central role in the production of abstract – homogenous, fragmented, and hierarchical – space. That is the space of capital. Lefebvre situates this socio-spatial mediation at the heart of the survival of capitalism.[39] This is also the core of his critique of bourgeois urbanism. Bourgeois urbanism is "capitalism's and the state's strategic instrument for the manipulation of fragmented urban reality and the production of controlled space."[40] His spatializing of the state and state-like provides a nuanced spatial dimension to Marxist theories of uneven development and offers the first explicit theorization of hegemony as a spatial project.[41]

Lefebvre's focus on the role of the state in organizing the relations of center-periphery at various geographical scales and historical moments eventually culminated in his concept of "colonization,"[42] which is a useful concept for analyzing the role of state urban strategies in pacification processes. "Colonization" refers to the political organization of territorial relations of domination,[43] once these relations are mobilized as a productive force in determining accumulation strategies, dynamics of displacement, and forms of spatial organization.[44] For Lefebvre center-periphery relations of domination and accumulation are not only found in the former colonies, but they are "extend[ed] to the heart of metropoles."[45]

Reading Neocleous and Lefebvre together directs us to how pacification is central to 'colonial' state urban strategies both in the metropoles and colonies. In fact, by the mid-nineteenth century, the commitment to sustain the empire shifted the attention of segments of the (colonial) ruling class to the disciplinary, ideological, and economic values of spatial strategies. It was not accidental that the late nineteenth and early twentieth centuries shaped one of the revolutionary conjunctures for military strategy, architecture, and urbanism. Arriving in Algeria in 1840, the leading French theorist of colonial war, Thomas-Robert Bugeaud,[46] professed rapid military offensives – a combination of total destruction and construction – as the main spatial strategy for French conquest and pacification.[47] Appropriating the *razzia* (raid) from the pre-Islamic Bedouin tribes, Bugeaud put into practice what today is known as urbicide: destroying the (built) environment to kill resistance and insurgency.[48] In the aftermath of 1848, Bugeaud's spatial strategies of pacification in Algiers inspired Baron Hussmann's re-organization of central Paris. To quell any possible revolution, Hussmann built the grand boulevards for the Parisian bourgeoisie on the ruins of working-class districts.[49] These were the same boulevards that the working-class revolutionaries re-appropriated during the Commune of 1871.

The micro spatial relations of occupation that Bugeaud put into practice were linked to the macro spatial-strategic relations of French colonization, namely, France's colonial policy of annexation and assimilation of the colonies.[50] By the end of the nineteenth century, however, important segments of the French colonial ruling class gradually replaced (albeit not as a whole) the assimilation strategy of the direct colonial rule with the Protectorate and the strategy of association. These changes influenced not only the French colonial conquest and pacification strategies for years to come but also the pacification doctrines of France and the United States after the WWII.[51] Fundamental to the ideology of association was an attempt to articulate colonization as a "humanist" project.[52] Consequently, indirect rule (association), the *politique des races*, and construction (rather than total destruction à la Bugeaud) became the outmost effective strategies for "humanizing" pacification at this conjuncture. The *politique des races* was meant to give French colonization a "native" human face by letting the colonized rule themselves through loyal local clients. The continuation of the *politique des races* can be traced in the emphasis on "participation" in the US domestic and imperial pacification strategies in the 1960s.[53]

The popularity of the *politique des races* and indirect rule as supposedly "humanist" ways of pacifying, dominating, and exploiting the colonized alert us to how pacification is mediated by struggles over spatial organization *and* the racial dimensions of the spatial relations of domination. Lefebvre's concept of "colonization" and Neocleous's concept of "pacification" thus need to be linked to Frantz Fanon's spatiotemporal character of colonization[54] and his critique of the colonial politics of recognition.[55] Fanon's nuanced analysis of colonization as a "spatial organization" and of everyday racism as "alienating spatial relation"[56] reminds us that any analysis of space in the colonial context should take into consideration the spatially mediated colonial relations of domination. The colonial spatial relations of domination, as Fanon neatly depicts, are not only integral to the political economy of capitalist-colonial exploitation and accumulation, but are also embedded in everyday racism, humiliation and violence that shape the subjectivities of the colonizer and the colonized. In this context, the work of Lyautey is of crucial importance.

HUBERT LYAUTEY: COLONIAL PACIFICATION AND URBANISM

The military and intellectual theories and practices of French colonial figures such as Thomas Bugeaud (in Algeria), Joseph Gallieni (in Tonkin and Madagascar), and Louis-Hubert Lyautey (in Madagascar and Morocco) were imperative to the formation of colonial warfare in the late nineteenth and the early twentieth centuries.[57] Among these figures, Lyautey, later France's first Resident-General in Morocco, became the most famous of France's colonial figures of the time. His pacification legacies are still studied by imperial armies of our times.[58] Lyautey was more than a colonial officer/warrior, however. He was a reformist public figure who frequented the highest Parisian literary and social circles[59] and considered himself an amateur urbanist.[60] Lyautey's military thinking should be situated in the context of the counter-revolutionary social-reform revolution of the nineteenth and early twentieth centuries. Two important spheres of this reform revolution were architecture and colonial warfare. The reform movements in architecture and urban planning came to assist the spatial regulation of everyday life with the aim of producing a healthy, efficient, and productive social order and subjects.[61] Amazed with the disciplinary power of spatial regulation, groups of colonial architects and governors became interested in experimenting with their top-down social-reform ideas as a modality of pacification in the colonies. In many ways, Lyautey was the frontrunner in making the link between urbanism and warfare a necessity for colonial rule.

Lyautey's theory of pacification was born, in part, out of his critique of the French army of the Third Republic.[62] To save France from decline, sterility, and revolution, Layutey proposed social pacification (with the goal of national unity). In his 1891 article, *Du rôle social de l'officier*, he argued that the true role of the army was not war but education and the true vocation of the modern army was social peace.[63] Thanks to his colonial experiences, he soon extended his concept of social pacification of France to the colonies, building upon the then slowly popularized theory of association. As an enthusiastic supporter of imperial France, Lyautey thought of the colonial empire as "the most solid element of power and growth" essential for "the smooth operation" of French national economy.[64] Over time his Orientalist fascination with the colonies resulted in "humanizing" his support for imperialism, arguing that imperialism is morally justified only if it contributes to the economic well-being and happiness of all humankind.[65] It was this "humanized" conception of empire that shaped his pacification theory, in which a combination of modern "improvement"[66] and respect for "traditional ways of life" was mobilized as a force to quell potential revolts.[67]

The colonial experiences most influential to developing Lyautey's pacification theory were his assignments to Indochina in 1894 and to Madagascar in 1900, where he combined and extended the colonial strategies of Governor-General Jean-Marie de Lanessan and Colonel Joseph Gallieni. In Indochina, De Lanessan was influential in putting a neo-Lamarkian scientific theory of association on the radar of the French colonial establishment, forcefully arguing that indirect rule over colonies was the best way to establish quick, effective, and inexpensive colonial control.[68] Colonial indirect rule was heavily dependent on spatial and socio-political knowledge of the colonized spaces, as each colony had to be organized on the basis of scientifically-established particularities of climate, customs, and political organization.[69] Lyautey also took great lessons from the spatial and territorial dimensions of Gallieni's strategies of *politique des races* and *tache d'huile*. "[T]he triumph of the Gallieni's method," for Lyautey, was based on linking together colonial rule and conquest in a way that "the latter [would have] no other objective than the former."[70]

In Madagascar, Gallieni consolidated his strategies of *politique des races* and *tache d'huile*. The pacification of Madagascar took the appearance of a developmental agenda that saw the natives improve through better markets, education, and hygiene.[71] At the heart of his work was a holy trinity: roads, schools, and markets. This trinity was central to regulating the everyday life of the colonized. In reality, the outcomes of Gallieni's pacification were systematic large-scale expropriation of land, the production of right-less, disciplined labourers, and the transformation of Madagascar into a modern state. He did so through changing the territorial, social, and property relations, strategic town planning, and the enforcement of *politique des races* under the administration of French military officers.[72] The political reorganization of territorial relations of domination meant that Madagascar was quickly compartmentalized into provinces, districts, and *fokon'olona*,[73] each with different relations to French colonial rule. Similar to the pacification of Algeria under Bugeaud, the new built environment in Madagascar during this period of pacification directs us to what Lefebvre understands as the relation between the micro spatial relations of occupation and the macro spatial-strategic relations of colonization. The construction of new buildings, as Gwen Wright demonstrates in detail, "represented the associationist approach to colonization."[74] Many of these were not simply new kinds of buildings, rather they were new spaces and institutions fundamental for the political administration and pacification of society, such as: clinics, nurseries, orphanages, post offices, town halls, and *maisons de tous*, as well as jails and police stations.

The principles of colonial urban design that Lyautey famously put into practice a decade later in Morocco emerged gradually during the pacification of Madagascar. Between 1900 and 1902, while responsible for pacifying the south of the island, Lyautey "established his first true *ville nouvelle* adjacent to the Malagasy city of Fianarantsoa. With a population of only 5000 people, this was already the second largest settlement in Madagascar."[75] Once in Paris in 1900, he summarized his colonial experience in Madagascar in a second article, entitled '*Du rôle colonial de l'armée*,'[76] extending his argument on the social role of the army to the colonies. The essay was an attempt to counter the anti-militarism of those segments of French ruling class who rejected both the high costs of military-colonial expansion and the violence of military rule in the colonies.[77] This idea of the colonial role of the army was to be perfected in Morocco.

The socio-spatial transformation of Morocco under Lyautey's indirect rule is a powerful example of how the spatially mediated colonial relations of domination were imperative to simultaneous materialization and reification of pacification strategies. As in Madagascar, for the purpose of political administration, Morocco was compartmentalized into various territorial units (regions, territories, *cercles* and *bureaux*) and scales (central, regional and municipal). Implementing the *politique des races* meant that most of the *pashas*, *qa'ids*, and their *khalifas* were initially left in their pre-colonial posts. They were, however, stripped of most of their prerogatives and powers and eventually replaced by agents of the newly formed *Services Municipaux*. One of Lyautey's first acts was to reorganize territorial relations of domination (Lefebvre) by establishing a centralized urban order across Morocco. Marginalizing the hitherto system of several regional royal capitals, this new urban order was based upon two political and economic poles: the cities of Rabat (the administrative and cultural capital) and Casablanca (a commercial and industrial capital), both situated on the coast, both looking toward Europe.[78] The geographical relocation of the capital from inland Fez to coastal Rabat was imperative to French accumulation strategies. It was based on favouring "the development of the major coastal cities tied to the French economy and the extraction of wealth from the inland regions."[79] A dual urban-regional system eventually resulted in the systematization of uneven development in Morocco.

As Janet Abu-Lughod has well documented, technically, legally, and spatially, Lyautey's pacification strategies established the infrastructure for the (colonial) capitalist state in Morocco.[80] At the same time, the elaborate system of laws and regulations, "gave a patina of legitimacy and equity to the process" of colonial urban planning, state formation, and pacification.[81] Urbanism was central to the materialization and reification of Lyautey's theory of pacification in Morocco. Two years after the establishment of the Protectorate, in 1914, the first comprehensive urban-planning legislation in the French world was decreed in Morocco.[82] The legislation heralded urban planning as a "public good" and called for master plans for all cities in Morocco. Upon transferring the capital to Rabat, a year into the formation of the protectorate, in 1913, Lyautey appointed the young French architect of the *Beaux Arts*, Henry Prost, as the chief planner of Morocco. Prost's task was to provide Lyautey with "a well-ordered, logical city plan, adapted to local conditions."[83] During his ten years leading urban planning in Morocco, Prost drew up master plans for nine cities: Casablanca, Rabat, Fez, Marrakesh, Meknes, Sefrou, Ouezzane, Taza, and Agadir. This colonial urban vision fundamentally transformed property relations and spatial relations of domination in Morocco.

Lyautey and Prost were both keen in mobilizing state power for utilizing spatial strategies as a means for reorganizing socio-spatial relations while keeping intact the racialized (that is colonial) dimensions of spatial relations of domination and accumulation. Despite an appearance of respect for the natives through his *politique des races* and of emphasis on the colonized participation, Lyautey was a strong believer in paternalism, centralized power, and top-down intervention.[84] State power was the condition of any coherent planning in Prost's architectural thought and practice.[85] Lyautey and Prost were influential in justifying the need for new laws that gave the French colonial administrative apparatus the police power to expropriate land for public purpose.[86] New property laws paved the way for the large-scale privatization and commodification of land through either selling public land to French investors and settlers by auction or expropriating public land by the municipality and then reselling it at a profit to Frenchmen for building the French *villes nouvelles*.[87] The result was a massive and rapid expropriation of land and the eventual formation of a form of spatial apartheid between the Moroccan cities and the French *villes nouvelles*. Two decades after the 1914 urban-planning legislation, by 1937, the French not only had registered more acreages than Moroccans, but also held title to lands whose value far exceeded that of Moroccans.[88]

If urbanism was influential to this forceful reorganization of property relations and socio-spatial relations of domination, it was also fundamental to moderating the violence of these processes of accumulation by dispossession. Adorning his project of pacification-via-urbanism with a colonial politics of recognition (Fanon), Lyautey in practice appropriated a form of liberal "humanist urbanism" that appeared to respect Moroccan "culture" and "way of life." A spatial-racial ideology of separation and preservation was at the heart of Lyautey's "humanist urbanism" and Morocco's urban-planning legislation of 1914. One dimension of this ideology was to extend the *politique des races* into the space of the colonized by showing respect for the aesthetics and social integrity of Moroccan cities.[89] This was called the strategy of preservation of the *medinas*. The second dimension was the application of the modern principle of urbanism in an ambitious program of constructing segregated *villes nouvelles* for the European settlers adjacent to Morocco's older cities. Both of these ideologies were already present in Prost's architectural work by the time he got to Morocco. Central to his conception of city and regional planning, for example, was a juxtaposition of the traditional and the modern, based on the principle of separating cultures in the name of respect for difference.[90]

"Yes," Lyautey famously declared, "in Morocco, and it is our honour, we conserve." And not just conserve: "I would go a step further, we rescue."[91] This rescue mission-cum-pacification hegemonized an aestheticized politics that was deeply spatialized and racialized. The spatial-racial ideology of separation-preservation resulted in a hierarchical spatial separation of the colonized and the colonizer in Morocco, what Abu-Lughod called "urban apartheid."[92] The same ideology simultaneously solidified a colonial politics of recognition (Fanon) that reified the racial dimension of such alienating spatial strategies and the resultant everyday racism in the name of "humanism," "diversity" and respect for "ways of life." While the divide between the colonized and the colonizer was more porous than Abu-Lughod would allow for it,[93] in practice, preservation designs sealed Moroccan cultures and ways of life within an imagined Orientalist past. And while Moroccans were deemed to be frozen, both politically and economically, at an archaic level of the picturesque, their (cheap) labour was exploited to develop the *villes nouvelles* for French colonizers.[94] The aestheticized preservation of the *medinas* was also part of Lyautey's conscious effort to commodify the everyday life and spaces of the Orient (Moroccan) by promoting a colonial tourism industry in Morocco.[95] For Prost, urbanism had a core aesthetic dimension. The imperative of the aestheticization of urbanism (and politics) was so high that Prost imposed aesthetic laws to control the development of Moroccan cities.[96] Lyautey's appropriation of a modality of "humanist urbanism" as a pacification strategy is perhaps best exemplified in his praise of Prost:

> The art and science of urbanism, so flourishing during the Classical Age, seems to have suffered a total eclipse since the Second Empire. Urbanism: the art and science of developing human agglomerations, under Prost's hand is coming back to life. Prost is the guardian, in this mechanical age, of humanism. Prost worked not only on things but on men, different types of men, to whom *la Cite* owes something more than roads, canals, sewers and a transport system.[97]

Lyautey's pacification strategies were not immune to contradictions and failures, of course. Eventually, in 1925, the Rif Rebellion brought an end to his dream of pacification. Lyautey's ideas, however, lived longer than his own ruling time-space in Morocco. On the one hand, the legacy of Lyautey and Prost in Morocco consolidated an ideology of racial-spatial separation/segregation that influenced colonial urbanism for years to come and has been transplanted across the colonial and metropolitan geographies throughout the twentieth and the twenty-first centuries. On the other hand, it resulted in a functional-romantic style, which Francois Beguin has aptly named "Arabisances."[98] Discussions about architecture and urbanism in Morocco were often translated into proposals for reform in France and her colonies. A few years after his failure in Morocco, Lyautey was appointed as the commissioner of the 1931 Colonial Exposition in Paris, where his ideas and practices influenced the International Congress on Urbanism in the Colonies and the subsequent new planning rules in French colonies.[99] Advising readers of *L'Architecture d'aujoud'hui* in the same year, Lyautey[100] advocated for applying his Moroccan formula to France: "Conserve the traditional fabric of Paris," "but outside of Paris be as daring as possible,"[101] anticipating the gentrification of central cities more than half a century earlier. The double ideology of separation-preservation that Lyautey put into an aestheticized and racialized practice in Morocco deeply influenced all later urban planning decisions in Algiers.[102] For example, on the eve of the War of Independence in Algeria in 1953, Chevallier, the French mayor of Algiers at the time, followed in the footsteps of Lyautey by adopting a form of "humanist urbanism"[103] aimed at constructing housing projects as a way of nullifying the anti-colonial revolution.[104] An similar attempt to Hussmann's a century earlier was miserably defeated, as many of those housing projects turned into the hotbeds of the anti-colonial revolution in Algiers.

RETHINKING URBANISM AND PACIFICATION

What lessons can we take from Lyautey's legacy in urbanism and pacification to have a better understanding of the reemergence of pacification in our conjuncture? Throughout the chapter my focus was on the centrality of "conceived space"—that is the space conceptualized by the ideologies of the state, planners, and urbanists – in the specific context of the French colonial pacification and urbanism. Lyautey's "humanist urbanism," I argued, directs us to the ways that the pacification of the local population (who were deemed dangerous *and* exotic) was mediated through the racial dimensions of the multi-scalar (re)production of hierarchized, fragmented, and homogenized space (Lefebvre) of Morocco. His *politique des races* mobilized a liberal-colonial politics of recognition that facilitated and reified the participation of the colonized in the processes of pacification. Ironically, both the partial success and the failure of Lyautey's "humanist urbanism" was based on his attempt to moderate violence and to reify the racial dimensions of domination by positing colonial pacification as a colonial politics of recognition of the colonized "culture" and "way of life" (Fanon). This urbanization of pacification was crucial to the "subjective dimension of the colonial experience" (Cesaire). As Fanon forcefully reminds us, the resultant "restricted picture of the world that the racialization of life and thought" produced, foisted on the colonized "the sever constriction of the space and shapes the colonized's moral and even political consciousness."[105]

I conclude by suggesting that Lyautey's "humanist urbanism" directs us to the ways that today the state mobilizes urban policy (and politics) as a modality of pacification for sustaining the hegemony of the imperialist capitalist order at various moments and scales. As mentioned earlier, the shift toward a more 'humanitarian' way of war and development is not limited to the zones of neo-colonial occupations and extractions. Parallel ideological shifts have been influential in re-defining the strategies of state-bound intervention in the urban 'priority zones' at the heart of imperial metropoles, where the majority of the non-White working class population reside. Herein, 'humanist' discourses of 'social mixity' and 'revitalization' are mobilized to moderate the violence of the state-bound strategies of redevelopment, gentrification, and policing that in practice are aimed at dispersing, fragmenting, and disciplining working-class populations, nullifying the possibilities of revolutionary politics.

Conceiving urbanism as a potential strategy for pacification that aims to reorder the relations of force can deepen our understanding of how, in our 'post-colonial' times, state-bound strategies of urban intervention and policing function as a modality of (neo-)colonial pacification. Such a project, nonetheless, needs more nuanced analysis of the processes and contradictions of state spatial strategies. It requires extending our analysis beyond the "perceived space" of such projects, as Foucauldians put it, to include the dialectical triad of the production of space.[106]

NOTES

1 I would like to thank Colleen Bell, Stefan Kipfer, Mark Neocleous, Liette Gilbert, Tyler Wall, Ryan Toews, and the anonymous reviewers at Red Quill Books for their comments and suggestions on the earlier drafts of this chapter.
2 See, John Nagl, *Learning to Eat Soup with a Knife: Counterinsurgency Lessons from Malaya and Vietnam* (Chicago: University of Chicago Press, 2005); David Kilcullen, *The Accidental Guerrilla: Fighting Small Wars in the Midst of a Large One* (London: C. Hurst & Co., 2009); Robert M. Cassidy, *Counterinsurgency and the Global War on Terror: Military Culture and Irregular War* (Westport, CT: Praeger Security International, 2006); Thomas Hammes, "War Evolves into the Fourth Generation," *Contemporary Security Policy* 26, no. 2 (2005): 189-221; Rupert Smith, *The Utility of Force* (London: Allen Lane, 2005); John Mackinlay, *Defeating Complex Insurgencies* (London: Royal United Services, 2005).
3 See Nils Gilman, "Preface: Militarism and Humanitarianism," *Humanity* 3, no. 2 (2012): 173-178; Feichtinger, Mortiz, S. Malinowski, and C. Richards, "Transformative Invasions: Western Post-9/11 Counterinsurgency and the Lessons of Colonialism," *Humanity* 3, no. 1 (2012): 35-63; Colleen Bell, "Civilianising Warfare: Ways of War and Peace in Modern Counterinsurgency," *Journal of International Relations and Development* 14 (2011): 309-332; Mark Neocleous, "War as Peace, Peace as Pacification," *Radical Philosophy* 159 (2010): 8-17; Conor Foley, *The Thin Blue Line: How Humanitarianism Went to War* (New York: Verso, 2010).

4   Military strategists have been increasingly mining lessons from the so-called 'classic' counterinsurgency literature–i.e. Thomas Edward Lawrence's *Seven Pillars of Wisdom: A Triumph* (1917), David Galula's *Counterinsurgency Warfare: Theory and Practice* (1964), Roger Tinquier's *Modern Warfare: A French View of Counterinsurgency* (1964), and Sir Robert Thompson's *Defeating Communist Insurgency: Experiences From Malaya and Vietnam* (1966).
5   For some major works, see endnote 28.
6   Douglas Porch, "Bugeaud, Gallieni, Lyautey: The Development of French Colonial Warfare," in *Makers of Modern Strategy*, ed. Peter Paret (Princeton: Princeton University Press, 1986); Douglas Porch, *The Conquest of Morocco* (New York: Knopf Inc., 1982); Thomas Rid, "The Nineteenth Century Origins of Counterinsurgency Doctrine," *Journal of Strategic Studies* 33, no. 5 (2010): 727-758; Thomas Rid, "Razzia: A Turning Point in Modern Strategy," *Terrorism and Political Violence* 21, no. 4 (2009): 617-635; Etienne De Durand, "France," in *Understanding Counterinsurgency: Doctrine, operations and challenges*, eds. T. Rid and T. Keaney (London: Routledge, 2010); J. Gotman, "Bugeaud"; Moshe Gershovinch, *French Military Rule in Morocco: Colonialism and its Consequences* (London: Frank Cass, 2000); W. A. Hoisington, *Lyautey and the French Conquest of Morocco* (New York: St. Martin's Press, 1995).
7   Gwen Wright, "Tradition in the Service of Modernity: Architecture and Urbanism in French Colonial Policy, 1900-1930," *The Journal of Modern History* 59, no. 2 (1987): 291-316; *The Politics of Design in French Colonial Urbanism* (Chicago: University of Chicago Press, 1991); Janet Abu-Lughod, *Rabat: Urban Apartheid in Morocco* (Princeton: Princeton University Press, 1980); Zeynep Celik, *Urban Forms and Colonial Confrontations: Algiers under French Rule* (Los Angeles: University of California Press, 1997); Paul Rabinow, "Colonialism, Modernity: the French in Morocco," in *Forms of Dominance: On the Architecture and Urbanism of the Colonial Enterprise*, ed. N. AlSayyad (Aldershot: Avebury, 1992); *French Modern: Norms and Forms of the Social Environment* (Cambridge, MA: MIT Press, 1989).
8   Jean-Jacques Deluz, *Urbanisme et l'architecture d'Alger* (1988).
9   Henri Lefebvre, *De l'etat II* (Paris: Union Generale d'editions,1976).
10 Gilman, "Preface"; Bell, "Civilianising Warfare"; Eyval Weizman, *The Least of all Possible Evils: Humanitarian Violence from Ardent to Gaza* (New York: Verso, 2011); Foley, *The Thin Blue Line*; Mark Duffield, "The Liberal Way of Development and the Development-security Impasse: Exploring the Global Life-chance Divide," *Security Studies* 41, no. 1 (2010): 53-76; Mark Duffield, "Global Civil War: The Non-insured, International Containment and Post-interventionary Society," *Journal of Refugee Studies* 21, no. 2 (2008): 145-165; Mark Duffield, *Development, Security and Unending War: Governing the World of People* (London: Polity Press, 2007).
11 Michael Cowen and Robert W. Shenton, *Doctrines of Development* (London and New York: Routledge, 1996).
12 Bell, "Civilianising."
13 Michael Dillon and Julian Reid, *The Liberal Way of War: Killing to Make Life Live* (Abingdon: Routledge, 2009).
14 Such dis-articulation is uneven throughout the Foucaldian and post-structuralist studies. Most often, 'global liberal governance' comes across as a code word for imperialist capitalism. Colleen Bell, "Civilianising Warfare," for example, situates the contemporary military and security policies in relation to imperialism. While Michael Dillon and Julian Reid, *The Liberal Way of War*, look at the political economy of liberal rule and war, they, nonetheless, take an explicitly anti-Marxist approach.
15 Neil Smith, *American Empire: Roosevelt's Geographers and the Prelude to Globalization* (Berkeley: University of California Press, 2004); David Harvey, *The New Imperialism* (Oxford: Oxford University Press, 2003).
16 Harvey, *The New Imperialism*, 184.
17 Jerome Klassen and Greg Albo, eds., *Empire's Ally: Canada and the War in Afghanistan* (Toronto: University of Toronto Press, 2013); Todd Gordon, *Imperialist Canada* (Winnipeg: Arbeiter Ring Publishing, 2010); Ellen Meikins Wood, *Empire of Capital* (London: Verso, 2003); Harvey, *The New Imperialism*.
18 Karl Marx and Friedrich Engels, *The Communist Manifesto (*New York: Penguin Books, ([1848] 2002); Friedrich Engels, *The Conditions of the Working Class in England* (Oxford: Oxford University Press, [1845] 1992).

19 Stuart Elden, *Terror and Territory: The Spatial Extent of Sovereignty* (Minneapolis: University of Minnesota Press, 2009); Simon Dalby, "Regions, Strategies and Empire in the Global War on Terror," *Geopolitics* 12, no. 4 (2007): 586-606; Beate Jahn, "Barbarian Thoughts: Imperialism in the Philosophy of John Stuart Mill," *Review of International Studies* 1, no. 3 (2005): 599-618; Uday Singh Mehta, *Liberalism and Empire* (Chicago, IL: University of Chicago Press, 1999); Cowen and Shenton, *Doctrines of Development*.
20 Domenico Losurdo, *Liberalism: A Counter-history*, trans. Gregory Elliot (New York: Verso, 2011); Neocleous, *War Power*; Neocleous, "The Dream of Pacification."
21 Robert D. Kaplan, "The Coming Anarchy: How Security, Crime, Overpopulation, and Disease are Rapidly Destroying the Social Fabric of our Planet," *The Atlantic*, February 1994, 44-76; Mary Kaldor, *New and Old Wars: Organized Violence in a Global Ear* (Cambridge: Polity, 1999); Mary Kaldor and S.D. Beebe, *The Ultimate Weapon is no Weapon: Human Security and the New Rules of War and Peace* (New York: Public Affairs, 2010); Thomas Barnett, *The Pentagon's New Map: War and Peace in the Twenty-first Century* (New York: Berkley, 2004).
22 This fact has been overlooked in critical security and IR studies as well as in urban and critical geography research.
23 Loic Wacquant, *Urban Outcasts: A Comparative Sociology of Advanced Marginality* (Cambridge, UK: Polity, 2008).
24 Anthony Gramsci, *Selections from the Prison Notebooks* (London: Lawrence and Wishart, 1971), 180-185. For the spatial dimension of Gramsci's analysis of the relations of force see Stefan Kipfer and Gillian Hart, "Translating Gramsci in the Current Conjuncture" in *Gramsci: Space, Nature, Politics*, eds. Mike Ekers, Stefan Kipfer, and Alex Loftus (Sussex: Wiley-Blackwell, 2013), 323-344. Gillian Hart, "Denaturalizing Dispossession: Critical Ethnography in the Age of Resurgent Imperialism," *Antipode* (2006): 977-1004.
25 Neocleous, *War power*; Neocleous, "The Dream," 11-12.
26 Neocleous, "The Dream," 31.
27 Neocleous, *War Power*; Neocleous "War as peace"; On police project see, Neocleous, *The Fabrication of Social Order: A Critical Theory of Police Power* (London: Pluto Press, 2000).
28 For the spatial dimension capitalism see Henri Lefebvre, *State, Space, World: Selected Essays*, eds. Neal Brenner and Stuart Elden (Minneapolis: University of Minnesota Press, 2009); Henri Lefebvre, *The Production of Space*; The Urban Revolution (Minneapolis: University of Minnesota Press, 2003); Henri Lefebvre, *The Survival of Capitalism: Reproduction of the Relations of Production*. trans. Frank Bryant (London: Allison & Busby, 1976). Harvey, *New Imperialism*; *Paris, Capital of Modernity* (London: Routledge, 2003); Harvey, *Spaces of Capital: Towards a Critical Geography* (London: Routledge: 2001). For the spatial dimension of colonization see, Anthony King, *Urbanism, Colonialism and the World-economy: Cultural and Spatial Foundations of the World Urban System* (London: Routledge, 1990); *Colonial Urban Development: Culture, Social Power and Environment* (London: Routledge and Kegan Paul, 1976); Abu-Lughod, *Rabat*; Zeynep Celik, *Empire, Architecture and the City: French-Ottaman Encounter, 1830-1914* (Washington, DC: University of Washington Press, 2008); Zeynep Celik, *Urban Forms*; Zeynep Celik, "New Approaches to the 'Non-western' City," *Journal of the Society of Architectural Historians* 58, no. 3 (1999): 374-381; Stephen Legg, *Spaces of Colonialism: Dehli's Urban Governmentalities* (Oxford: Blackwell, 2007); M. Fuller, *Moderns Abroad: Italian Colonial Architecture and Urbanism* (London: Routledge, 2006); Manu Goswami, *Producing India: From Colonial Economy to National Space* (Chicago: Chicago University Press, 2004); For the spatial dimension of war and security see, Jon Coaffee, *Terrorism, Risk and the Global City: Towards Urban Resilience* (Surrey, UK: Ashgate, 2009); Stephen Graham, *Cities Under Siege: The New Military Urbanism* (New York: Verso, 2010); Stephen Graham, ed., *Cities, War, and Terrorism: Towards an Urban Geopolitics* (Malden: Blackwell Publishing, 2004); Derek Gregory, *The Colonial Present: Afghanistan, Palestine, Iraq* (Oxford: Blackwell, 2004); Eyval Wizeman, *Hollow Land: Israel's Architecture of Occupation* (New York and London: Verso, 2007); Eyval Wizeman, *The Least of All Possible Evils*.
29 Neocleous, "The Dream," 18.
30 T. Todorov, *The Conquest of America: The Question of the Other*, trans. Richard Howard (Oklahoma: University of Oklahoma Press, 1999).
31 Losurdo, *Liberalism*; Neocleous, *War Power*.
32 King, *Urbanism*, 29.
33 Henri Lefebvre, *The Urban Revolution* (Minneapolis: University of Minnesota Press, 2003); Henri Lefebvre, *The Production*, 150-151, emphasis added.

34 King, *Colonial Urban Development*; King *Urbanism*.
35 The spatialization of Marxism from the 1960s to the late 1970s directed attention to the role of urbanization in the survival of capitalism. Since the late 1980s, influenced by Michael Foucault, the dominant theme of the literature on colonial urbanism has increasingly become the spatialization of (colonial) knowledge and power. Among the many 'links' that shaped such dialectical formation are: 1) the systematic appropriation of the colonies as laboratories for colonial architectures and planner; 2) the immensely important function of colonies in the realm of culture; 3) the expression of political power in urban design and space; and 4) how spatial organization and eventually urban planning have been fundamental to the disciplinary and formative nature of modern society. See endnote 28 for a selected list of these works.
36 One of the main shortcomings of this literature is that colonialism is mainly understood as a (cultural) encounter, rather than also a systematic feature of the geographical expansion of capitalism. As such, the complexities of domination have been mainly approached through an emphasis on 'modernity' and 'tradition.' For an historical-materialist critique of understanding colonial relations of domination solely as a cultural encounter see, Himani Bannerji, *Democracy and Demography: Essays on Nationalism, Gender and Ideology* (Toronto: Canadian Scholar's Press, 2011).
37 Frantz Fanon, *The Wretched of the Earth* (New York: Grove Press, [1963]2004); A Sekyi-Out, *Fanon's Dialectic of Experience* (Cambridge: Harvard University, 1996); Stefan Kipfer, "The Times and Space of (de-)colonization: Fanon's Countercolonialism, Then and Now," in *Living Fanon: Global perspectives*, ed. Neal Gibson (New York: Palgrave Macmillan, 2011), 93-104.
38 For an in-depth debate see, Stefan Kipfer, Parastou Saberi, and Thorben Wieditz, "Herni Lefebvre: Debates and Controversies," *Progress in Human Geography* 37, no. 1 (2013): 115-134.
39 Lefebvre, *The Survival*, 97.
40 Ibid., 13.
41 Stefan Kipfer, "City, Country, Hegemony: Antonio Gramsci's Spatial Historicism," in *Gramsci: Space, Nature, Politics*, eds. Mike Ekers, Gillian Hart, Stefan Kipfer, and Alex Loftus (Sussex: Wiley-Blackwell, 2013), 83-103; Stefan Kipfer, "How Lefebvre Urbanized Gramsci: Hegemony, Everyday Life and Difference," in *Space, Difference, Everyday life: Reading Henri Lefebvre*, eds. Kanishka Goonewardena, Stefan Kipfer, Richard Milgrom, and Christian Schmid (New York: Routledge, 2008), 193-211.
42 For an in-depth discussion see, Stefan Kipfer and Kanishka Goonewardena, "Urban Marxism and the Post-colonial Question: Henri Lefebvre and 'Colonisation'," *Historical Materialism* 21, 2 (2013): 76-116.
43 Kipfer and Goonewarndena, "Urban Marxism," 95.
44 Lefebvre, *De l'etat*, 173-174; cited in Kipfer and Goonewardena, "Urban Marxism," 95.
45 Kipfer and Goonewardena, "Urban Marxism," 95.
46 Thomas Bugeaud, *La Guerre des Rues et des Maisons* (Paris: J-P Rocher, [1849] 1997).
47 Porch, "Bugeaud,"; Rid, "The Nineteenth Century"; Rid, "Razzia"; Anthony Thrall Sullivan, *Thomas-Robert Begeaud, France and Algeria 1784-1849: Politics, Power, and the Good Society* (Hamden: Conn, 1983).
48 Martin Coward, "Urbicide in Bosnia," in *Cities, War and Terrorism*, ed. Stephen Graham (Malden: Blackwell Publishing, 2004), 154-171; Martin Shaw, "New Wars of the City: Relationships of 'Urbicide' and 'genocide,' " *Cities, War and Terrorism*, ed. Stephen Graham (Malden: Blackwell Publishing, 2004), 141-153; Stephen Graham, "Constructing Urbicide by Bulldozer in the Occupied Territories," in *Cities, War and Terrorism*, ed. S Graham (Malden: Blackwell Publishing, 2004), 192-213.
49 Bugeaud returned to Paris from Algiers in 1847 and published the treatise *La guerre des rues et des maisons*, which is known as the first manual for urban warfare. As Eyal Weizman has mentioned, Begeaud understood that the relationship between the organization of the space and the ability of the military to control the space. See Weizman, "Military Operations as Urban Planning," *Metamute*, 2003, and Sharon Rotbard, *White City, Black City* (Tel Aviv: Babel, 2005).
50 Gershovich, *French Military Rule*, 20.
51 Laleh Khalili, *Time in the Shadows: Confinement in Counterinsurgency* (Stanford, CA: Stanford University Press, 2013), 23.
52 See Wright, *The Politics*; Rabinow, *French Modern*.
53 See Stuart Schrader, in this volume.
54 See Kipfer, "The Times"; Skeyi-Out, *Fanon's Dialectics*.

55 Glen Coulthard, *Red Skin, White Masks: Rejecting the Colonial Politics of Recognition* (Minneapolis: University of Minnesota Press, 2014).
56 Kipfer, "The Times and Spaces," 94.
57 Porch, "Bugeaud," 376-377.
58 See Rid, "The Nineteenth Century."
59 See Rabinow, *French Modern*.
60 Lyautey's interest to associate himself with urbanism was so deep that a recurrent element in his biographies, as Paul Rabinow highlights, "is the depiction of the young Lyautey kneeling in his garden, assiduously building cities (not chateaux) in his sand pile, joining his father's engineering talents with the artistic flair of a precocious urbanist." See Rabinow, *French Modern*, 107, and Wright, *The Politics of Design*, 8.
61 Rabinow, *French Modern*.
62 Rabinow, *French Modern*; Porch, "Bugeaud"; Hoisington, *Lyautey*.
63 Rabinow, *French Modern*, 123.
64 Cited in Hoisington, *Lyautey*, 10.
65 Rabinow, *French Modern*, 151.
66 For the appropriation of 'improvement' in pacification strategies see Neocleous, *War Power*. Also see Will Jackson and Helen Monk in this volume.
67 Wright, *The Politics*.
68 Hoisington, *Lyautey*, 6.
69 Rabinow, *French Modern*, 138.
70 Cited in Hoisington, *Lyautey*, 8.
71 Wright, *The Politics of Design*; Khalili, *Time in the Shadows*.
72 Wright, *The Politics*.
73 Rabinow, *French Modern*, 159-162. *Fokon'olona* literally means 'grouping of persons.' Paul Rabinow translates it as *deme*. George Condominas defines it as territorially-based patrilocal clans localized around an ancestor's tomb and providing religious unity. As in Vietnam, the *fokon'olona* was first identified as the French sense of municipal community. Lyautey instituted legislation which conceived of the *fokon'olona* as a territorial unit rather than a kinship or ritual grouping.
74 Wright, *The Politics*, 254-255.
75 Wright, *The Politics*, 259.
76 The article was published in the *Revue des Deux Mondes*.
77 Hoisington, *Lyautey*, 16.
78 Wright, *The Politics*, 95.
79 Wright, *The Politics*, 95.
80 Abu-Lughod, *Rabat*.
81 Abu-Lughod, *Rabat*, 147. This legal system drew extensively from recent urban legislation in Germany, Italy, Switzerland, and Egypt, see Wright, *The Politics*, 351, footnote 115.
82 Morocco's planning legislation preceded by almost five years the passage of similar (though not nearly so far-reaching) legislation with respect to French cities, and by two years the famous New York City Zoning Resolution.
83 Rabinow, *French Modern*, 301.
84 Rejecting equality, Lyautey substituted a paternalistic *demo-philie*: "I love the people, among whom I live. But if I love with all my heart, it is as a protector (*patron*) and not a democrat." Cited in Rabinow, *French Modern*, 283.
85 Rabinow, *French Modern*, 238.
86 Abu-Lughod, *Rabat*, 174-195.
87 Abu-Lughod, *Rabat*, 163-169.
88 Abu-Lughod, *Rabat*, 170-171.
89 Rabinow, *French Modern*, 288.
90 Rabinow, *French Modern*, 255.
91 Cited in Celik, *Urban Forms*, 40.
92 Abu-Lughod, *Rabat*.
93 See Celik, "New Approaches."
94 Abu-Lughod, *Rabat*, 171-173.
95 Abu Lughod, *Rabat*; Wright, *The Politics*.
96 Called *police de construction*, these regulations did not directly legislate style, but they did set rigorous guidelines for scale, materials, services or alignment in various districts, see Wright, *The Politics of Design*, 05.
97 Cited in Rabinow, *French Modern*, 288-289.
98 Wright, *The Politics*, 108.

99 Celik, *Urban Forms*, 40.
100 Hubert Lyautey, "Architecture", *L'Architecture d'aujoud'hui* 1, June-July (1931): 14.
101 Cited in Wright, *The Politics of Design*, 137.
102 Celik, *Urban Forms*, 39.
103 Deluz, *Urbanisme*.
104 Celik, *Urban Forms*, 143.
105 Ato Sekyi-out, "Fanon and the Possibility of Postcolonial Critical Imagination," in *Living Fanon: Global Perspectives*, ed. Neal Gibson (New York: Palgrave Macmillan, 2011), 51.
106 Lefebvre, *The Production*.

Fortification and the Fabrication of Colonial Labour

Aaron Henry

As Mark Neocleous has noted, military strategists are returning to primary military documents and texts in a bid to revitalize pacification for the twenty-first century.[1] This chapter takes up the call that critical scholars should do likewise and re-examine the history of pacification so as to uncover the practices and tactics that have made up the 'relations of force' in bourgeois society. The relationship of pacification to the production of urban space is a persistent theme throughout this collection. The chapters by Christopher McMichael, Parastou Saberi, and Philippe Zourgane direct our attention to pacification and the spatiality of counter-insurgency, war, 'urban planning,' and the destruction and creation of new landscapes. Saberi's study of the early twenty-century urban planning and counter insurgency strategies of Louis-Hubert Lyautey elegantly suggests that pairing pacification to space reveals "the ways the state can mobilize urban policy and politics as spatial strategies of pacification."[2] Following this line of reasoning, I use this chapter to interrogate the colonial strategies of rule and fabrications of productive labour that have preceded pacification and the production of urban space. I focus this study on the different ways the Hudson's Bay Company (HBC) designed and deployed forts throughout the eighteenth and nineteenth centuries in British North America.

Why study strategies of pacification through eighteenth and nineteenth century Hudson's Bay Company forts? In some ways the fort is a rather obvious site of examination as it is a persistent feature of European projects of conquest and is more or less coextensive with the rise of pacification in the sixteenth century. Yet, despite the rich empirical history to be culled from the fort's long career in organizing colonial relations of rule, it has remained, for the most part, the possession of military historians. Yet, as Stephen Webb has suggested, historically the fort and the city were identical—forts were cities and cities were forts.[3] If this is true, the spatial history of the fort may serve to reveal the longstanding entwinement of war and colonial urban space in projects of pacification.

To this end, I develop two arguments. First, I argue that by comparing eighteenth and nineteenth-century fort designs we can locate a shift in the fort from an enclosed site of mercantile extraction to a site of inspection charged with making indigenous people into productive labour.[4] Second, I argue that from the mid-eighteenth to early nineteenth century, the fort shifted from being conceived of as singular headquarters of power and instead became imbricated in part of a network deployed to develop an extensive spatial field of rule. I suggest that if it is the case that pacification is linked with the production of new spaces, urban space and 'programmed landscapes,' the shift in the design and deployment of the fort reveals how this link is historical and constituted in projects of developing administrative landscapes that are capable of rendering labour productive.

This chapter proceeds in three parts. First, I sketch how the fort was imagined as 'an art of enclosure' in the eighteenth century by examining the work of John Muller, a professor of fortification and artillery at the Royal Military Academic of Woolwich (1699-1784). I then discuss how some of Muller's maxims on fortification are reflected in the HBC fort diagrams from this period. Second, I discuss how from the end of the eighteenth century and throughout the nineteenth century the fort was restructured and deployed in service of a new administrative environment. I chart this shift in two ways. First, I provide a discussion of how the HBC opened up forts, reduced the scale and scope of the walls, and also added a new building called the 'Indian Hall' that gave the company a greater ability to keep records of indigenous hunters. Second, I point to how this architectural innovation coincided with attempts by Sir George Simpson, the governor of Rupert's Land (1821-1860) to make indigenous labour more productive by employing better systems of registration to identify indolent and 'infirm' indigenous hunters. I point to how these two changes inserted the fort into a new and more extensive administrative landscape. In the final section, I conclude on what this can tell us about the spatial history of pacification.

JOHN MULLER AND THE EIGHTEENTH CENTURY ART OF ENCLOSURE

John Muller was a professor of artillery and fortification at Woolwich University from 1736-1766. Trained as mathematician, he first wrote on geometric principles before applying these principles to warfare in a series of treatises on military strategy and fortification. For the purposes of this discussion, I focus on Muller's "A Treatise on Fortification" published in 1756. Muller opens this text with the declaration that 'the fort' is an "art of enclosing towns" and "commanding space." And in this sense, forts may be used "to guard some passes, in a mountainous country or near causeways, rivers and other such places."[5] To control the 'entrance into a country' and therein to control the country itself and its land, resources and subjects, is a matter of correct enclosure. The right size of fort, the right walls, in the right location lends command over the country and regions surrounding the fort. One rules over what is without from within.

Parallel to the fort's rationalization of space, Muller advanced the fort as a site of commerce. "The greatest strength of [England]" wrote Muller "and all maritime nations consists in traffic." And, as "fortified places which lie near rivers, lakes, creeks, or the sea chiefly serve to protect and promote trade" the construction of such types of forts must be explained "in a particular manner [so that] nothing [will be] omitted."[6] If one builds a fort on the river large enough to house a market, "the streets are to be perpendicular to the river, and to cross each other at right angles" so as to conveniently facilitate the "transporting goods from the ships to the market."[7] In addition, in selecting a site for construction, the prince or governor must know the surrounding country. In particular, is "the country round plentiful in those things which are necessary for the sustenance of men and cattle?" And, above all, "it [must] be considered whether the country round it produces commodities fit for trade [and] how [they] can be improved and where [they] may be transported to, either for sale or exchange."[8] As the "great wealth of a country depends on traffic by sea" one must build forts that have "proper landing places" and the capability to "defend against the force or stratagem of any enemy." The fort must have "good magazines, convenient storehouses to lodge the goods in and in general everything should be thought of that contributes to easy conveyance and the proper disposition of [commodities]."[9] The design and placement of eighteenth century Hudson's Bay Company forts appears to reflect Muller's maxims.[10]

The Hudson's Bay Company was granted the 'sole trade and commerce in the waterways and 'sea streights' of the Hudson Bay basin by royal charter in 1670.[11] From 1670 to 1763 the HBC consolidated its hold over the basin of the Hudson Bay through the construction of five forts: York Factory, Fort Albany, Fort Charles, Fort Rupert and Fort Prince of Wales. These forts were smaller models of European fortifications.[12] Placed no more than ten kilometers from sea or rivers, their geographical location betrays their purpose: food and building supplies could be exchanged for furs and shipped back to London. The image below is an HBC fort plan for Churchill dated in the mid-eighteenth century. The highlighted section reads "place for Indian trade."

However, despite these buildings being purposed with large storerooms to house the accumulation of goods, practices of exchange were not intended to disrupt the fort's enclosure. Throughout the eighteenth century trade at HBC forts was conducted along the exterior wall in the designated 'Indian trading places,' as highlighted in the image below. These locations were positioned near each fort's bastions in clear lines of fire. Some designs included 'trading wickets' or small portholes in the exterior palisade for goods to be exchanged.[20] In this sense, the fort's architecture ensured that the wall was only porous to goods, not men, women, and children. In line with Muller's advice on using the fort to control traffic and trade, the HBC fort was explicitly designed as a site of extraction. Its very structure and geographical location presupposed that furs and goods would be deposited by indigenous hunters, packed into storerooms, and then shipped back to London. In fact, these five forts formed the nexus for one of the primary relations of extraction in British North America. Thus, the eighteenth-century fort was shaped by mercantile relations of production: accumulation was premised on holding trade routes and traffic rather than through increasing the productivity of labour directly. To this end, the eighteenth-century fort presided over a limited social field. The palisades, the walls, all served to produce a circulation of goods that admitted commodities but ideally kept 'the natives' outside the fort's gates. There is an important point here that serves to contextualize the changes made to HBC forts' architecture and function in the nineteenth century.

The eighteenth-century fort, by virtue of the social relations of production it codified, belonged to a limited social field. Accumulation of wealth was expressed through correct geographical placement. The countryside must be selected that 'produces such commodities as are fit for trade' and that 'can be improved' and transported 'for sale and exchange.' Therefore, the fort must be planned in such a way that is supports the circulation of commodities and their 'improvement' before export. This economic reasoning bears a great deal of resemblance to the economic doctrines of the eighteenth-century police science.[21] The strategic placement of HBC forts along rivers and tributaries within reach of the sea reflects this logic. Thus, in the eighteenth century the HBC fort commanded what was 'without' from within. It is the placement of the fort that lends the prince or governor control over the 'entrances into the countryside'. Rivers, causeways, mountain passes, and ocean straights are the fort's principle objects of rule as each confer power over the traffic of goods. Consequently, besides ensuring that the immediate surroundings of the fort are 'wholesome,' what lay outside the fort was not subject to regular superintendence. In fact the 'art of enclosure' precludes this very possibility. The fort is deployed to ensure that a limited number of men control an entire country through the application of force to trade routes and lines of traffic.

Consequently, the eighteenth-century fort did not presume to command an extensive social field of people, land, and things. Reviell Netz has argued, in his seminal account of 'Barbed Wire,' up until the mid-nineteenth century, rule over land and people strikes the modern eye as 'strangely inverted'.[22] According to Netz, before the mid-nineteenth century trade routes and contact points were commanded from specific headquarters of power and, consequently, the rest of the social field remained unknown and outside control. Of interest to this discussion is that if part of pacification refers to a process of subjecting populations to regular surveillance and classification, the eighteenth-century fort, by virtue of its supreme enclosure, is incapable of developing this form of rule. As a consequence, the HBC was incapable of building knowledge of indigenous people as a population, a limitation that constrained its ability to compel them to be productive. This limitation is reflected in the company's systems of registration and documentation from this period.

FORTIFICATION AND THE FABRICATION OF COLONIAL LABOUR 111

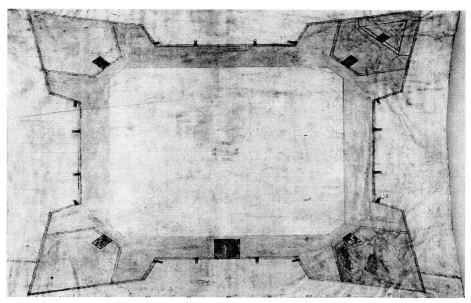

Above Design: Church Hill Plan Circa 1730[13]

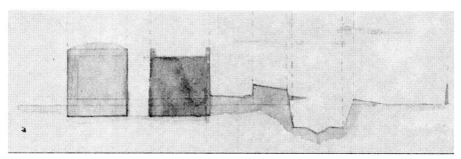

Design 2 and a sketch of the outer walls: defense begin with stakes, an open plain easily scoured by musket fire, a moat, an embankment with wooden stakes, an external wall, and finally the stone perimeter of the keep.[14]

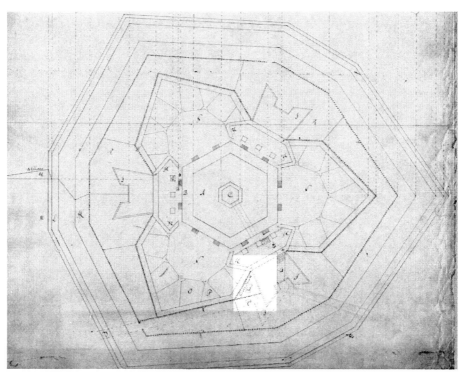

All of the HBC forts in this period had one entrance kept perpendicular to the entrance in the fort's keep. A sentry was stationed to watch the perimeter and ensure no HBC employees left the fort without the governor's permission.[15] All walls, bastions, canon batteries, and ramparts were planned according to geometric principles.[16] The exterior space of the fort, as per Muller's instructions, was divided into quadrants of angles that rendered the fort's immediate surroundings visible.[17] As one of the architects noted, the fort must "inclose a space of ground that… can be viewed by every side so there may be no shelter where any enemy can lodge themselves."[18] Thus, the fort was designed to bring its immediate area into permanent view so as to ensure that targets could be clearly identified and bombarded before enemy canon shells or fire created breaches. In the designs from this period walls were given primacy and the interior space of the fort was often left blank. The exception to this was the inclusion in draft plans of designated storerooms comprising no less than 8,000 square feet of storeroom space. This specification was necessary if the fort was to fulfill its function as a site of extraction.[19]

In 1749, parliament commissioned a select committee to inquire into the state of the HBC affairs. Company employees were asked to attest to whether the company had faithfully undertaken its civilizing duty towards its 'native subjects.' The company's response was that to civilize them would require the company "to bring them up to labour from their youth," something that was unequivocally impossible as the company knew little about the locations of these populations; they simply arrived every spring to trade.[23] Within this same period, the company's documentation of indigenous populations was notably vague. John Isham, a company servant for nearly 30 years, when asked to write on the 'native' population was only able to state that they "are not numerous relative to the proportion of land they occupy."[24] The same year Isham made this statement, the company asked William Coates to develop an account and map for a new Fort in the Gulf of Richmond. In his survey, Coates described the immediate features that would fall within the purview of the newly planned fort. Coates described the wealth in the trees, the beasts, the plains, and then remarked "of men: that we expected to see the footsteps of many numerous tribes of Indians which we did everywhere in our progress."[25] This lack of detailed knowledge about Indigenous people, their location, and their numbers appears to have persisted within the HBC all the way up until 1815.[26] It can be argued that this inability to gather information on indigenous people is structurally linked to the materiality of the fort itself. While the HBC produced meticulous accounts of goods that entered and exited the fort, the enclosed nature of the fort foreclosed on keeping records of indigenous people, despite the fact that they were the principal suppliers of the companies 'wealth.' Consequently, the registers of the eighteenth-century fort were incapable of fabricating a colonial labour force as an object of rule.

Although this structure of accumulation did not pose a problem for the company during the eighteenth century, by the start of the nineteenth century the inability to know and control the labour process of indigenous hunters manifested in the company's economic decline. While the decline in the company's fortunes was linked to a number of events, its loss of markets in Europe due to the Napoleonic wars and the depletion of beaver stock, one key contributing factor was an explosion of debt precipitated by an overextension of credit in HBC trading practices.[27] Extending goods (rifles, ammunition, clothes, cooking supplies) to indigenous people on credit was a key practice in the company's operations. Issued in the fall, these advances were intended to allow indigenous hunters to purchase guns, ammunition, clothing, and supplies. The company reasoned this ensured hunters spent less time procuring subsistence and more time hunting furs for the company. In its course, this relationship of social and economic reproduction increased the company's power over the hunters' means of subsistence and made them dependent on the HBC's credit.[28]

The standard practice had been to issue the value of 8 to 10 Made Beaver (MB) pelts at the start of the trade year. Following their hunt, indigenous hunters would have this advance deducted from their returns. However, without any system of registration it appears that it was difficult to accurately trace from year to year which advances had actually been paid. In fact, as late as 1816-1817, advances were still paid by taking an automatic deduction from hunters' returns rather than following any record. James Tate, placed in charge of Henley House, recorded his dissatisfaction with an inattentive young clerk for failing to deduct advances from a party of indigenous traders. The party informed Finlayson that they had already paid their all debts at another post. Finlayson 'believed them' and took their furs at the full rate, for which he was reprimanded. The fact that there was no debt book for Finlayson to consult suggests the company had yet to properly record and individualize debt.[29] The records available in the company's archives further indicate a lack of sustained documentation, as it is not until 1821 that abstracts of 'Indian debt' appear as a stable and consistent administrative document. I argue that it is context to the need to police debt and encourage hunters towards greater productivity that the fort transformed from being conceived by the company as a singular site of enclosure and became a site of inspection and registration that was imbricated in a far more extensive social field.

THE FORT IN THE NINETEENTH CENTURY: ARCHITECTURE, SURVEILLANCE, LABOUR

In 1786 an innovation arrived in the design of Fort York Factory. The change appears slight: the new design converted two existing rooms with a shared wall into a counting room and a trading room. The trade room had a single door and was lined by windows, three of which were outside and one inside linking it with the counting room. The architect noted in the design plan that "it might be deemed unsafe to admit too many of the natives at one time into the fort."[30] The new design admitted indigenous people inside the fort; however, their movement was to be carefully controlled and subject to a range of controls. The trade room, for instance, was placed close to the main gate and separated from the inner buildings and the fort's keep by another set of gates, effectively isolating the movement of hunters from the rest of the fort. Nonetheless, the wall of the fort was made porous to a planned circulation of indigenous hunters. Crucially, the trade room shared the same wall as the new counting house. Furs passed over one side, rifles, ammunition, blankets, tobacco, liquor, and cookery passed from the other.

The pairing of the clerk's office and the trade room was not merely an architectural episode, a dream of 'order' that remained unrealized. 1815 Floor plans show that the configuration was still in use.[31] Moreover, the architectural configuration proliferated and over the nineteenth century became a standard arrangement of most HBC forts. In fact, from 1815 onward the exterior wall was increasingly removed from design plans.[32] The careful coordination of walls and canons that had dominated the design plans of the eighteenth-century fort receded from the structural layout of fort or outpost. By the mid-nineteenth century one heavily fortified building, a keep, guarded by "high wooden stakes" was all that remained of the previous heavily fortified structure. In short, there was a push for an economy of power focused less on enclosure and more on the management of circulation both from within and outside the fort.

It appears that this push towards a lighter and more 'open fort' became a company standard under the direction of Sir George Simpson. Simpson, trained in the counting houses of London, was selected in 1821 by the HBC board of directors to revitalize the company's fortunes. Also in this year, the company amalgamated with long time rival, the North West Company. This merger not only inaugurated a number of changes in the company's accounting practices[33] but it also meant that the company was in possession of new territory (Alberta, British Columbia, Oregon, and Washington State) and in need of modernizing its forts and trading posts in these areas. We can get some insight on how the design from York Fort spread by examining Simpson's comments on Fort George during his 1824 Inspectorial Tour.

Formerly a North West Company (NWC) holding, the HBC established control over Fort George following the amalgamation of the two companies in 1821. The American Navy had initially built Fort George in 1811; however, within months of its completion indigenous people raided the fort and the inhabitants were killed.[34] The NWC took control of the fort one year later but, for obvious reasons, erected more palisades and strengthened the bastions. George Simpson described the fort as follows:

> The establishment of Fort George is a large pile of buildings covering about an acre of ground well stockaded and protected by Bastions or Blockhouses, having two eighteen Pounders mounted in front and altogether an air of appearance of Grandeur and *consequence which does not become and is not at all suitable to the Indian Fur Trade post*.[35]

It seems Simpson found the fort too fortified. The stockades, bastions, and weaponry that had made up the core features of HBC fort design plans only some 30 years earlier had become unpalatable and ill-suited to conducting 'the Indian fur trade.' Simpson's remarks are perhaps all the more curious given that many HBC servants continued to rely on the fort as a defensive structure. In 1823, for instance, a clerk in the Pembina District requested five additional men to guard the fort while repairs were underway to refortify it. The men were requested because one man had been "recently killed by Indians not far from the [fort's] stockades."[36] Thus, Simpson's criticisms did not stem from a sudden absence of opposition from Indigenous populations. We may glean some insight into why this structure had become unsuitable from Simpson's remarks on Fort Vancouver, the fort he commissioned shortly after his visit to Fort George.

> The fort is well picketed covering a space of about 3/4ths of an acre and the buildings already completed are a Dwelling house, two good stores, an *Indian Hall*, and temporary quarters for the people. It will in two years hence be the finest place in North American, indeed I have rarely seen a Gentlemen's seat in England possessing so many natural advantages and where ornament and use are so agreeable combined.[37]

There are two noteworthy features here. First, Simpson's fort has a lighter structure. It is still walled, but the multiple layers, glacis, and moats are reduced by the fort's strategic placement on a hillside. The elevation of the hill gives the fort a natural defense and a better firing position. In this sense, the fort was perhaps not only less costly but it also would not deter trade due to heavy fortifications. Second, Simpson is keen to list that the new fort will have an "Indian Hall." The "Indian Hall" became a regular design feature of HBC forts over the nineteenth century. An HBC Official recalls the place of the 'Indian Hall' in Mountain House outpost in 1850:

> Mountain house was surrounded by the usual 28-foot pickets, with a block bastion at each corner and a gallery running all round inside about four and a half feet from the top, each bastion containing a supply of flintlocks and ammunition … There were two gates, the main gate on the north and a smaller one on the south side leading a narrow passage the height of the stockade into a long hall. In this hall the Indians were received … They were then turned out and the gates closed against them, the only means of communication being through two portholes some 20 inches square opening through the stockade into small blockhouses through which the trade was conducted.[38]

Similarly, writing to Simpson in 1853, a chief factor described a new design plan for Fort Simpson. The fort was to be made of wood, 180 feet from the river, and all buildings are spaced by 50 feet. An 'Indian Hall' is included in the diagram. Like the earlier nineteenth century design at York Factory, the 'hall' was placed next to the exterior wall and given a separate entrance out of the fort.[39] The 'Indian Hall' codified a new arrangement of rule. The hall marked a movement away from the absolute restrictive space of the wall and instead recalibrated the fort to support a circulation of men, women, and children into the fort, albeit limited. Thus, the hall recast the fort as a site of containment where indigenous people were admitted and at once cloistered, confined to a site that posed little risk to the security of the rest of the fort. Yet, if 'the hall' allowed for the retrenchment of the wall, it also marked a further modification in power insofar as it introduced the possibility for new practices of registering indigenous hunters as labouring subjects. Thus, the 'Indian Hall' served to combine the registers of the counting house and the trade room into a singular list. It became the role of the clerk to account for who entered the fort, their name, their purpose, and record their trade. In this sense, the hall provided the material possibility to combine bodies into one site where they could be effectively recorded.

In a bid to revitalize the company's trade, Simpson implicitly enjoined the new architectural structure of the fort to a new system of administration. Upon taking power, Simpson moved away from assigning governors to specific forts and instead assigned chief factors to pre-selected trade districts. The district, a practice of division the company had been putting in place since 1815, extended the control of the HBC's trade away from individual forts to a collection of forts linked together in a pre-delimited administrative field. In addition to assigning company officials to a more extensive field of management, Simpson also decreed that chief factor and chief traders would be responsible for compiling an annual 'district report' for the company's records. While the report included the geography of the district, its rivers, the state of buildings, and the conduct of company servants, it also, notably, required chief factors to collect information about the 'Indian' populations in their respective districts. In particular, Simpson decreed that chief factors be directed annually to furnish registers of the number of Indians attached to their respective districts or parts particularizing the tribes, the number of chiefs and followers, *with extent of the country they inhabit and hunt in, together with their average debts given, returns brought to the company*, general character and habits of life.[40]

Simpson issued this request for two reasons. The first and most obvious was that the company wished to set a higher individualized debt limit for 'its' indigenous hunters. Thus, it was important to develop knowledge of debt and roughly their annual returns to see if this new debt limit was not too much of a liability.[41] Second, the request coincided with concerns from the company's directors in London over 'indolent populations.' At the end of the summer in 1822 Simpson received letters from the London Council expressing concerns over a growing unproductive population of deserted half-breed children and 'infirm' and' indolent natives' that frequented forts and outposts. The company argued that these populations 'were dangerous to the peace of the country and the safety of the trading posts' and needed to be removed to the Red River colony.[42] In this sense, it was expedient to get chief factors to keep registers of the debts of individual hunters and their returns, so as to locate which hunters were productive and which were 'indolent.' Thus, whereas the eighteenth-century fort had exerted very little control over indigenous hunters, the nineteenth century regime sought to build knowledge of colonial labour and root out the indolent. We can see how these processes worked in practice by examining some of the district reports produced by Simpson's chief factors. I will open here with Chief Factor William McRae's district report from the Flying Post district (1823-1829).

The first year that McRae was put in charge of the Flying Post district, he noted that his 'stay in this quarter has not been long enough to enable [him] to procure the information necessary to form a district report.'[43] He conceded that he had yet 'to visit in person the different trading posts' in the district and 'those in a subordinate station [clerks] had yet to furnish him with such information.' However, as he was 'unwilling to be silent altogether,' he consulted the records from the previous year at Flying Post and found there were 'forty Indian on the books' in the district.[44] All of them 'were very poor and the total debt of the district exceeds one thousand martins.' From this McRae concluded that Flying Post, the post he was stationed at in the district, ought to be closed, as the cost of maintaining it appeared too high compared to the other posts in the district.

The report for 1825-1826 opened with a review of the different posts in the district and a more thorough account of the district's gardens, the means of subsistence, the state of the buildings, returns of the trade, conduct of servants and an 'Indian abstract.'[45] McRae informed Simpson that the district's returns were poor. The cause was not only an unfavorable winter but also 'the yearly diminishing beaver and other animals in those places where the Indians are much too crowded and where each possess too small a portion of land to allow it alternate returns.' He elaborated that the decline in the district's profitability had been caused by disease: 'The Indians in the district suffered from an epidemical disease, which proved fatal to 5 aged persons, of whom two hunted annually to the amount of forty martins each.'[46]

McRae then reported the outstanding debt and industry of the 'Indians' in his district. What follows is an extract from his report. 'No. 3 This man hunted well formerly but is now old and decrepit, his hunt consequently diminishes yearly, he brought this spring but 36 martins.' 'No. 4, son to the above, wintered this year too near the house, his hunt is therefore not above half his usual quality.' An insertion is added, the 'foregoing Indians are … disposed to carry their furs to other [non-HBC] traders, induced to so by … the great differences in the price of goods.' No. 12 'brought but 22 martins which is not above half his usual amount.' No. 14 'brought 80 martins and the same quantity last year, maintained seasonal average.'[47]

McRae's district report for 1828 Flying Post District was augmented by a census. Titled 'Return of Population, Flying River District' the census form consists of the following categories. Males over the age of 15 are listed by name as hunters, with their owing balance or credit in an adjacent column. Next to this, a category is included for the number of the hunters' wives, their sons, under 15, and daughters. In providing a total, McRae is keen to demonstrate that in this district 25 hunters support a total population of 170. An additional column is made which includes the widows of the district and their total male and female children. Nine widows, three boys, and thirteen girls, take the population of the district up to 195. Using the knowledge gleaned from his 'return on population.' McRae notes that despite 'the death of five hunters and that one band traded almost nothing, there is no diminution of the amount of returns compared with last year.' This is because, noted McRae, 'young men growing up whose hunts are on the increase,' and given the number of young boys in the district 'a further increase particularly of Martin may be confidently expected' in the near future. At the same time, McRae pointed to 'how numerous the Indians are' and 'the small portion furnished by each hunter' to illustrate that the relative productivity of the hunters was overwhelmed by the district's large dependent population. In this sense, he advocated the district be closed due to the economic loss incurred by its operations.

The second example of the district report and extension of the fort is taken from the reports of Angus Bethune (1783-1858). Bethune, a holdover from the NWC, negotiated the amalgamation of the two companies in 1821. In the amalgamated company he was given the role of chief factor and put in charge of the Albany District.

Bethune opened his district report with an account of the principle furs in the district, followed by the wildlife, the gardens in cultivation, the state of the buildings, the character of HBC servants and the 'Indian hunters and their debt.' The report then divided the district into its respective forts and posts. As Bethune had admitted to 'having little personal knowledge of the establishments,' he included the narrative of post clerks into his report on the profits derived from each post and the state of the establishment. Yet, despite not having the knowledge of the particular topographical features of each establishment, Bethune was able to call to hand the number of hunters contained 'in [his] district.' He remarked that he had 54 hunters on the 'books.' The records indicate that their primary hunts were 'chiefly in martins' but that they 'they also kill a few beaver and otters.'[48] What is notable though is that Bethune refused to give an account of 'the character of each individual hunter at present.' His reasoning was that he remained unfamiliar with the district and that 'the season has been so extremely unfavorable to hunt that [he couldn't] form a correct opinion of their actual capacities.'[49]

With this caveat, Bethune relayed information about four notable hunters. He, like McRae, did not refer to them by name but instead listed them as the 'number' that they were entered as in the district debt book. 'No. 20. This man gave me amongst other furs forty large winter beaver, not one cub and thirty winter otters, he has made the best hunt of any.' 'No 1. And No. 39 are brothers, good hunters but presumed dead.' 'No. 7 and No. 8 traded furs for ammunition,' 'but, [Bethune suspects] are headed to Severen district to get more debt.' By the following year Bethune was confident enough to remark on the character of the 'Indian population.' In a section now labeled 'Indian Abstract,' Bethune listed the debts of 57 Indigenous hunter for the last three years. Each individual, entered once again as a number, was cross-referenced against a 'remarks section.'

Bethune goes on provide more details: 'No. 1, an old man who is unable to hunt much, he has become of late more infirm. Should be removed to a place with good fishing. Starved last winter.' 'No. 4 a young man can hunt well when he pleases very indolent should not get credit.' 'No. 7 a bad Indian but can hunt when he likes very well, should not get debt.' 'No. 11, a good hunter always exerts himself. When he fails paying his debt it cannot be attributed to indolence.' 'No. 17, a good indian getting old. In a good season is industrious.' 'No. 41 a young man a good hunter but has been unfortunate as many others were last winter.'[50]

Bethune also used the report to note his suspicions that the company's credit system was being 'defrauded' as hunters are travelling between districts and doubling up on debt. He wrote,

> Two of the Severn Indians came to me this spring with the intention of trading their furs and by that means defrauding their own trader ... This is the first instance they have had of our determination to put a stop to their practice of fraud and which was done in the presence of about 30 Indians [a show was made of it] ... Their intention was positively to get as many goods as they could from me and return to Severen to play the same game there ... Without a strict understanding between Severn and Albany the business of these districts would be carried at a loss.[51]

In response to Bethune's report, the company decreed that chief factors would 'interchange of indian debt with gentlemen in charge of neighbouring districts to prevent fraud.'[52] In addition to sharing information on the populations in their districts the company also decided that indigenous people would not be permitted to leave the districts to which 'the gentlemen [had] assigned them.'[53] It appears that this practice of confining Indigenous people to specific districts was to be enforced throughout the HBC's operations. By 1841 the company's project of holding the 'Indian population' in place by district had culminated in Simpson's plan, only ever partially realized, to organize districts by 'tribe,' giving each tribe its own delimited hunting grounds, a practice that appears curiously similar to the reservation system that would be initiated at the turn of the nineteenth century in Canada.[54] In both Bethune's and McRae's reports, indigenous people are not only enumerated on and made visible as a population but they are classified as either productive (industrious) or idle (indolent) labour. The company's request that chief factors compare the current returns to past seasons even effectively transformed the hunting of furs into a labour process that could be quantified and compared to some 'social average' in each district. This form of representation though required that indigenous hunters no longer remain outside the fort but instead be made subject to regular and extensive registrations during their seasonal exchanges, something that was clearly impossible under the architectural arrangement of the eighteenth-century fort. I want to conclude here by examining two major themes that I think this analysis reveals about pacification and its relation to the production of spatial order.

CONCLUSION

One of the core features of pacification projects is that they are pursued for the purpose of 'extending the scope of productive labour.'[55] Both in contemporary and historical projects this objective often relates to getting people who previously were subsisting outside wage-labour to become subjugated to capitalist productive relations. What this study of the HBC underscores is the unique spatiality of these processes in British North America. In particular, while Marx familiarized us with primitive accumulation and the historical process of producing 'free-labour' by dispossessing peasants by forcibly enclosing pastures, rivers and so on, the HBC reveals a different set of spatial logics in the colonial theatre.[56] Given the size of territory, the number of HBC staff and the mobility of indigenous people, direct enclosure at the start of the nineteenth century was unfeasible.[57] Instead, rather than directly enclosing space, the HBC's tactic was to control the mobility of people and confine them to distinct locations, which, in turn, impacted their capacity for subsistence and made them objects of colonial knowledge. Thus, the opening of the fort and the introduction of 'district reports' reveal the historical linkages between the objectification of labour and the historical emergence of 'programmed' or 'administrative' landscapes in colonial British North America.

As evidenced by Bethune's and McRae's reports, the opening of the fort and the use of district reports were crucial processes in giving the company the power to track and compare the returns made by indigenous hunters. It is only by at once individualizing hunters, comparing their annual returns, that the categories of 'industrious' and 'indolent' could be stabilized. In keeping these records, individual hunters who refused to exert themselves to their best efforts could be disciplined by restricting access to credit or by simply being removed from the district. To generate productive labour it was not enough to account for what entered and exited the fort. What was requisite were registrations of hunters staged to the scale of the district. In particular in order for debt to be policed and resistance to be countered it was essential that chief factors coordinate their records to keep indigenous hunters constrained to their pre-determined district of 'residence.' By doing so not only was the doubling up of debt rendered 'impossible' but also it was possible to ensure that their returns in any given season were the 'real measure' of their productivity. Thus, rendering labour productive and anticipating the attempts of indigenous hunters to circumvent the HBC's power over credit required the recalibration of the 'space of the fort' so as to support a more extensive social field. In this sense, the history of developing pacified spatial environments and productive labour is in British North America tied up with the history of the fort. While George Rigakos has discussed the relationship of architecture to productive labour in the work of William Petty,[58] the HBC helps to develop the relation between the control of space and the formation of productive labour in the colonial history of British North America. In particular, the HBC's shift from ruling through forts to districts directs our attention to how the pacification of labour is historically tied to the production of a far more extensive social field of administration with a distinct imaginary of how people ought to placed and monitored in their environment.

Related to this point, the HBC's attempts to extend the scope of productive labour by ruling over a more extensive social field also sheds some light on the fabrication of population in the colonial imaginary. As noted by Neocleous, part of the shift from conquest to pacification during the reign of Philip II of Spain included the 'pacification' of the land, which was achieved by gathering 'information about the various tribes, languages and divisions of the Indians in the province.'[59] In many ways the history of the HBC reveals the further sophistication of population in the shift from conquest to pacification.[60] In particular, one has with the extension and opening of the fort the development of a whole new system of registration developed at the scale of the district. Population is not only produced through practices of enumerating individual hunters and their families, but also through the development of an accompanying topography. There are two important points here. The first point is that the formation of a colonial population required policing the mobility of individual indigenous hunters. In particular, by holding indigenous people in the space of the district the HBC was not only able to keep track of debt and productivity but it was also possible to produce an 'accurate' enumeration of people as population. Provisionally, I think there is probably a rich thread here on pacification and the politicization of 'mobility' during the nineteenth century. Similar concerns are actually raised in England in 1839 in discussions on how to develop constabulary districts throughout all of England and Wales.[61]

This shift in the spatiality of rule over the eighteenth and nineteenth century ushered in by processes of pacification offers up some connective threads to political rule in liberal societies more generally. In particular, according to Foucault, up until the sixteenth century sovereignty was 'not exercised on things but first of all on a territory and *consequently* on the subjects that inhabited it.'[62] As such, over the seventeenth and eighteenth century, sovereign power was directed against spaces that had remained outside earlier configurations of rule; 'square, markets, roads, bridges, and rivers' emerged as objects of regulation, especially within police apparatuses.[63] Foucault leaves us with the impression that power shifted from the surface of 'things' to a form of government concerned with the relations between land, people, and things across a more extensive social field. This argument is intriguing but Foucault doesn't really tell us much about why this shift occurred.

It can be argued that the history of pacification offers some clarification. The districting of the fort arguably deals with a shift similar to that which Foucault is referring from ruling land to trying to rule an interior set of conditions. It can be argued that processes of pacification, in particular the development of new landscapes to render labour productive, were crucial in recalibrating political rule from commanding a dominion to ruling people through planned administrative fields. By inverting the fort from an insular site to a nodal point through which an entire social field could be registered, it was possible for the HBC to consider the populations in each district in relation to the means of subsistence, geography, boundaries of the district, flora, fauna, buildings, and rivers, all of which produce additional knowledge that could be used in administration. It is quite clear that the eighteenth-century fort was not geared towards a thorough and continuous registration of what lay outside the fort. Thus, it is arguable that the process of pacification may serve to explain why power shifted from operating on land, dominions, and principalities and began to be concerned with constructing knowledge of people and their environment across a more extensive social field. Creating new administrative landscapes that developed extensive systems of knowledge over people, land, and things was essential in rendering labour both productive and visible.

In closing, I think this discussion has not only drawn attention to the entangled nature of pacification and spatial control, but it has revealed the particular relations of force that structured colonial projects of making indigenous people into productive labour in early nineteenth century British North America. This contribution is in itself important, as it has been largely overlooked by critical scholarship.[64] Moreover, in developing this history I think this chapter has situated the 'fort' as an important site of historical transformation in the formation of capitalist social relations; and, to this end, it reveals the place of the fort in a long historical trajectory of commanding space or developing 'programmatic' landscapes, as other chapters in this collection have discussed. Moreover, what has been demonstrated here concretely is that the HBC's attempts to make indigenous labour productive coincided with reworking the fort from an enclosed site to point of registration in an extensive field of rule; and, as such, this chapter underscores how processes of pacification initiate systematic changes to the geographical expressions of power and control.

NOTES

1 Mark Neocleous, "Brighter Happier Life: Security as Pacification" in *Social and Legal Studies*, vol. 20. (2011): 205.
2 In this collection see chapters by Parastou Saberi and Christopher McMichael.
3 Stephen Webb, *The Governors-General: The English Army and the Definition of the Empire 1569-1681* (USA: University of North Carolina Press 1987)
4 Throughout this discussion I have used the term 'indigenous' people. I would have liked to have been more specific and provided the reader with the actual names of the people that the HBC attempted to rule; however, the manner in which the company identified and documented the indigenous hunters who worked under them has made this quite difficult. Seldom are the nations of indigenous hunters identified in the HBC's documents. Moreover, when a collective name was provided, it was often based on the purported geographical location the company assigned to indigenous people. As a consequence, this chapter approaches indigenous people largely through the categories the company supplied. While I am uncomfortable with this, it does serve to illustrate the durability of colonial categories of rule.

5   John, Muller, *A Treatise Containing the Elementary Part of Fortification, Regular and Irregular* (printed in London, reprinted by the Museum Restoration Service, Ottawa, 1968 [1756]), 196.
6   Muller, *Treatise*, X.
7   Muller, *Treatise*, 172.
8   Muller, *Treatise*, 184-185.
9   Muller, *Treatise*, 187.
10 A direct relationship between Muller's design plans and HBC forts is difficult to find. However, most of his advice on angles, building material, and correct geographical positioning is replicated in "An Essay on the Timber Buildings in Hudson Bay" written for the HBC council by an English fort designer in the mid-eighteenth century. See AM 11M1 41/74 "An Essay on the Timber Buildings in Hudson Bay" date and author unknown, circa 1740s. The other line that connects Muller with fortification in North America is his work on artillery, which references and advises on military strategies for 'the Indian Wars'. See John Muller, *A Treatise of Artillery* (London: Whitehall, 1768).
11 Hudson Bay Company Charter, 1670, 1.
12 Kenneth Perry, *Frontier Forts and Posts of the Hudson Bay Company: During the Fur Trade and Gold Rush Period* (Canada: Hancock House Publishers, 2007).
13 Archives of Manitoba, (AM) 11M1/G1/79, Six Plans of Churchill not executed [undated Circa 1740].
14 AM, 11M1/G1/72, "Design of Fortification" [undated circa 1700].
15 David Mackay, *The Honorable Company* (United States: Bobbs-Merril Press, 1936), 220.
16 AM, 11M1/G1/80, "Geometric Plans of Fort Churchill" [circa 1760].
17 AM, 11M1, G1/78 Six Plans of Churchill.
18 AM, 11M1 G1/75, Six Plans of Churchill.
19 11M1 G1/41/74, "An Essay on the Timber Buildings in Hudson's Bay," [date circa 1740].
20 Mackay, *The Honorable Company*, 220.
21 See Johann von Justi, "Selections From on Staatwirthschaft," in *A General Police System: Political Economy and Security in the Age of Enlightenment*, eds. George Rigakos, John L. McMullan, Joshua, Johnson, and Gulden Ozcan (Ottawa: Red Quill Books, 2009).
22 Reviell Netz, *Barbed Wire: An Ecology of Modernity* (Connecticut: Wesleyan University Press, 2004), 63.
23 Cited in Edward Cavanagh, "A Company with Sovereignty and Subjects of its Own? The Case of the Hudson's Bay Company, 1670-1763," *The Canadian Journal of Law and Society* 26, no. 1 (2011): 37.
24 James Isham, *Observations on Hudson's Bay* (London: Hudson's Bay Record Society, 1949 [1743]), 104.
25 AM, 11M1/G1/17, "A Map of Artuimipeg or the Gulf of Richmond," William Coates, 1749.
26 A clerk at Whale's Lake Outpost was asked by the company to provide a complete enumeration of the native population including hunters, women, and children around the outpost. He conceded, "a list of those who trade at this settlement together with the number of their wives and children I have endeavored to collect as exact as possible but their are some who at times are not seen here for several successive years and whose names are mostly unknown to me" in AM, 1M778/B372/2/e/1, Report to Thomas Vincent Governor of the State of the Honorable Hudson Bay Company, Settlement at Whale River, Thomas Adler, 1815. [Sic]
27 See George Colpitts, "Accounting for Environmental Degradation in Hudson's Bay Company Fur Trade Journals and Accounting," *British Journal of Canadian Studies* 19, no. 1, 2006: 1-32.
28 Christopher Hanks, "The Swampy Cree and the Hudson's Bay Company," *Ethnohistory* 29. no. 2 (1982): 107; Arthur, Ray, *Indians in the Fur Trade: Their Role as Hunter, Trappers, and Middlemen in the Lands Southwest of Hudson Bay, 1660-1870* (Toronto: University Press, 1974), 117.
29 AM, 1M779/B/117/e/1, Henley House Report, James Tate, 1816-1817.
30 AM, 11M1, G1/112 "A Design for Building a Commodious Fort in Room of the Present Warehouses at Fort York Factory" (1785-1786).
31 1M11/G1/113.
32 AM 3M102/D5/36 AM, 3M102/D5/36, Plan of Fort Simpson Letter From James Anderson to George Simpson [1853]. The 'hall' was also included as a feature in the 1815 design plans for the Fort Wedderburn—AM 1M153/B235/a/3, "Rough Sketch plan of Fort Wedderburn" 1814-1815.

33 Colpitts, "Accounting for Environmental Degradation," 1-32, and Deidre Simmons, *Keepers of the Record: The History of the Hudson Bay Company Archives* (Canada McGill-Queen's University Press, 2009), 96.
34 See Perry *Frontier Forts and Posts of the Hudson Bay Company*, 10.
35 Cited in Mackay, *The Honorable Company*, 192.
36 AM, 1M814/B239/k1, Minutes of a Council Held at York Factory Northern Department of Rupert's Land", 1822.
37 Mackay, *The Honorable Company*, 193 [my italics].
38 Mackay, *The Honorable Company*, 229-230.
39 AM, 3M102/D5/36, Plan of Fort Simpson Letter From James Anderson to George Simpson [1853].
40 AM 3M43/D/4/87 [my italics]) "Minutes on the Assiniboia District" [1823-1824].
41 Hanks, Christopher, "Swamp Cree and the Hudson's Bay Company at Oxford House."
42 AM, 1M814/ B239/k/1, Northern Department Minutes of Council 1821-1831, Minutes of a Temporary Council, August 12, 1822.
43 AM, 1M778/B70.e/1, Flying Post District Report McRae, 1823-1824, William McRae.
44 AM, 1M778/B70.e/1, Flying Post District Report McRae, 1823-1824.
45 AM (1M778/B70/e/4) Flying Post District Report, 1825-1826, William McRae.
46 AM (1M778/B70/e/4) Flying Post District Report, 1825-1826, William McRae.
47 AM 1M778/B70/e/4, Flying Post District Report, 1825-1826, William McRae.
48 AM (1M776, B/3/E/8) Report from the Albany Department, Angus Bethune, 1823-1824.
49 AM (1M776, B/3/E/8) "Albany Report on District" 1822-1823, Angus Bethune.
50 AM, 1M776/B/3/e/9, "Albany Report on District" 1822-1823, Angus Bethune.
51 AM (1M776, B/3/E/8). Report from the Albany Department, Angus Bethune.
52 AM (1M814/B239/K1) "Minutes of a Council Held at York Factory Northern Department of Rupert's Land, July 8, 1823.
53 AM, 1M814/ B239/K2, Minutes of the Northern Department Council, July 5 1823.
54 For further description of Simpson's plan to organize 'families' by district see Arthur Ray, "Some Conservation Schemes of the Hudson Bay Company, 1821-50: An Examination of the Problems of Resource Management in the Fur Trade," *The Journal of Historical Geography* 1, no. 1 (1975): 61.
55 George Rigakos, " 'To Extend the Scope of Productive Labour': Pacification as a Police Project," in *Anti-Security*, eds. Mark Neocleous and George S. Rigakos (Ottawa: Red Quill Press, 2011), 72.
56 Marx Karl, "Chapter Twenty-seven, the Expropriation of the Agricultural Population" in *Capital vol. 1* (England: Penguin Press, 1991), 877-895.
57 For a discussion of the dispossession of indigenous people from their lands in Canada in the late 1880s onward see R. Cole Harris, *Making Native Space* (Canada: UBC Press, 2003). The scale of these projects and the use of state practices of documentation reveal the material impossibility for the HBC to realize a similar project of accumulation by dispossession. In this sense, the early nineteenth century marks a distinct form of primitive accumulation shaped by the structural limits of the HBC's capacity to rule.
58 See George Rigakos and Richard Hadden, "Crime, Capitalism and Risk Society: Towards the Same Olde Modernity," *Theoretical Criminology* 5, no. 1 (2001), 61-84.
59 Cited in Neocleous, "A Brighter and Nicer New Life: Security as Pacification," 199.
60 See Gulden Ozcan and George Rigakos, "Pacification," in *The Wiley-Blackwell Encyclopedia of Globalization,* 1st edition, ed. George Ritzer (Blackwell Publishing 2014), 2.
61 See *First Report of the Commissioners: Appointed to Inquire As t Best Means of Establishing an Efficient Constabulary Force in the Counties of England and Wales*, "Migratory Depredators" (London: Clowes and Sons, 1839), 13.
62 Michel Foucault, *Security, Territory, Population* (London, New York: Palgrave Macmillan, 2007), 96 [*my italics*]).
63 Pasquale, Pasquino, "Theatrum Politicum: The Genealogy of Capital—Police and the State of Prosperity," in *The Foucault Effect Studies in Governmentality,* eds. Graham Burchell, Colin Gordon, and Peter Miller (Chicago: University of Chicago Press, 1991), 111.
64 Harris has presented one of most robust analysis of colonial dispossession. However, he overlooks the historical period developed in this study by dismissing it as a period dominated by commercial capital and barter relations between HBC officials and Indigenous labour. See R. Cole Harris, "How Did Colonialism Dispossess? Comments from an Edge of Empire," *Annals of the Association of American Geographers* 94, no. 1 (2004): 165-182.

Programming the Landscape
Pacification Through Landscape Management

Philippe Zourgane

P ictures are scrolling on my computer screen. I'm looking at them in a distracted way. It is a long series of photos newly put online by the French Army visual archives service on its website. Most of them are photos taken during the decolonization wars. None of them are of real interest, only military parades, official visits of military or civilian figures, some civil engineering operations … I am looking for a series of photos of the Algerian vineyard planted by the French colonists.

I am caught by one series of agricultural works photos that seems to take place in Alsace or in the Beauce plain. There may be an error in the localisation of those harvest scenes in which the harvest is done by the soldiers themselves. They must have been taken near one military camp in France. One of the last photos frames a cameraman who is filming a sack full of grains around which are posted a French soldier and an agricultural worker. On the bags it is written in big letters ALGERIA. I have the confirmation: we are in presence of Algerian landscapes in the previous pictures.

Agriculture is often considered as a low-tech technology. The recent debates on biotechnologies and GMO remind us that, on the contrary, it has always been at the cutting edge of technology. Yet it is not often seen as a repressive technology apart from its initial economic purpose that is inducing the creation of a workforce and optimized exploitation of the land. James Scott says, "Concentrated population and grain production was, as we have seen, normally a necessary condition for state formation. Precisely because such areas offered a potential surplus for state building rulers, they also represented an irresistible target for raiders."[1] If large landscape management is first of all an economic history, it is also a political history nesting the machinery of State and economy.

Algeria was a French colonial territory from 1830 to 1962. This transformation of the landscape has been implemented very strongly in different ways. Beyond the management of urban centres there is the management of the colonial territory at various scales, from village units to bigger territorial units up to the global scale of the whole colony. Those three scales of simultaneous transformations were identified by the French Army as a tool for colonial war (Bugeaud, Gallieni, Lyautey). The Army is not only a conquest force but it is also a long-term occupying force whose task is to transform the dominated population. Pacification establishes subjection through the landscape itself.

FORBIDDEN ZONES

In Algeria, the Forbidden Zones appeared in 1954, as Fabien Sacriste describes:

> From November 1954, the civil and military authorities of the Aurès declared an area that included most of the Aurèsian Douars closed to human presence and traffic in its entirety and required the population to relocate to the principal cities of the region. The recourse to this strategy continued in the Aurès during the year 1955, without any overall doctrine being defined: a doctrine appeared for the first time in the directive of General Parlange, the Indigenous Affairs Officer of Morocco, appointed Military and Civilian Commander of Aurès-Nemenchas (CCMAN) in May 1955.[2]

This would become generalized, and the Forbidden Zones would spread in all the regions of Algeria, until the end of the war.

General Bugeaud, general governor of Algeria from 1840 to 1847, used the forbidden zones as a modus operandi for the confiscation of Algerians' lands. From the very beginning, pacification is linked to agricultural colonization: cutting down the olive trees, burning the orange fields and the wheat fields, sequestering the agricultural lands. On the one hand, the very violent military repression that is taking place is destroying and burning the villages in a systematic way. On the other hand, this military campaign aims to construct new farm units working especially for export. The transformation of native farmers into homeless people, the future proletarians in the service of big landowners, is crucial in this first phase. The logic of pacification put in place consists of destruction and construction to manage land and to secure European and local migrations.

The territory concerned is cleared, and it is rendered transparent. The residents themselves are 'clustered' into new, more distant buildings, near the existing urban units or in regroupment camps created from scratch, and kept under the surveillance of the army. The indigenous people are relocated into an 'urban milieu,' and the architecture will undertake to civilise them by providing them with the advances of the modern home. This is what Pierre Bourdieu called "the depeasanisation of Algeria."

After having been entirely occupied, the whole colonial space at one point has to be emptied, neutralised, sterilised. To start with, one possibility is to recreate white zones, but this time, they'll function differently. They won't anymore be white, designating the virginity of the conquered area of the future, a zone omitted from what's known, from control, a zone to be discovered, that could perhaps be tomorrow's new Eldorado, once colonised and then to be developed. This time, the affected area is to be emptied—of its residents, its herds, its cultivated land—to bring this zone back to its status of virgin land, to control it better, to transform it into a forbidden zone.

The only way to describe the goal is the sterilisation of landscape. The work to be implemented consists of preventing the use by the guerrilla of a space considered impossible to monitor beforehand as it would be if it were under administration. Concretely, it must be cleared[3] of its population in order to be able to control it afterwards. If this 'technique' was used several times during periods of crisis as, for example, in Malaysia by the British army, or in Vietnam by the French army in its "pacification" operations against the rebels, Communists, and terrorists, the process was intensified during the creation of Forbidden Zones in the Algerian War. These Forbidden Zones were at once military dimension but also contained a powerful psychological dimension linked together through the construction of the territory.

Only the colonial military are present intermittently, assuring the sterilisation of the space that is completely emptied of all livestock and human beings. "Pasture areas, as well as forbidden zones, are defined by straight lines that aren't materialized on the site, sometimes shepherds and more often camels are crossing those lines and are at once strafed by the airforce who often patrol in this area which is closed to Colomb-Béchar and Hammaguir air bases."[4] There remain only wild animals whose presence is controlled under the eyes of the military. Any 'otherness' has disappeared from the Forbidden Zone. Pierre Vidal-Naquet defines the Forbidden Zones as the areas "where you shoot everything that moves." Clear, Regroup, Monitor, and Shoot—taking control of this large landscape is effectively a prodigious efficiency for the state itself in its affirmation. The transformation of what Gilles Deleuze and Félix Guattari name "smooth spaces" "allows not only to defeat nomadism but also to control migrations and more generally to assert a zone of right on an 'exterior' ... it is vital for each state."[5] Space itself is modelled by geometry (the geometry of territorial or urban delineation, new road infrastructures) and organised by administration staff members (engineers, civil engineering services, architects). The deployment of state apparatus power transforms the territory. This is a military strategy to take back control of sparsely populated areas and put them under its administration. The Forbidden Zone is easy to monitor by helicopter or by other aerial means, 'legally' shooting any moving targets within that zone by strafing them from airplanes.

Here's the rationale put in place to eliminate the National Liberation Front (FLN): Clear the space of residents, of inhabitants, and thereby empty the space of terrorists and their logistical support. To get to that point, it was necessary to remove the population of all the villages and transfer them, to separate the FLN from the population and to protect the population from the FLN at the same time. As David Galula explains: "It soon became obvious that military operations alone could not defeat the rebels. The population had to be protected, controlled, won over, and thus isolated from the rebels. Work in depth was necessary. This is how we had pacified Morocco in the 1920s and early 1930s." This work—of 'managing' the population—was the consequence of that management of the countryside, of a vast territory that led in fact to the forced displacement and deportation of the population.

The evacuated villages were most often destroyed so they couldn't serve as refuges for the FLN, according to the French authorities, but also to prevent the return of the residents and to reorganise the territory with no traces of the past. That strategy was based on the counter-revolutionary doctrines developed during the experience in Indochina. During this period, a new generation of officers came to the command posts, which changed many things. The French army had lost the war in Indochina, but it didn't want to lose the war in Algeria. These new strategists of the French army had understood the teachings of Mao Tse-tung and of Ho Chi Minh. They were also watching the British in places like Malaysia where the Communist party and rebellion were destroyed.

Algeria was to be the colonial space where the French experience of managing a population[6] was successfully put in place, and where no countervailing power would be possibly exercised. Algeria was to be a police operation par excellence and at the heart of the operation were the Forbidden Zones. Entire populations were displaced arbitrarily and placed in camps that fabricated a new urbanism of hyper-visibility. And all these new techniques—or old techniques used on a new scale—were a success. Only the weight of international politics rendered this politic of repression untenable, because on the ground, it was a military success.

THE 1000 VILLAGES OF DELOUVRIER

These transfers affected millions of people (although no precise figures exist, and they are all subject to controversy). The first phase of the operations, directly orchestrated by the army, was succeeded by the second phase, implemented by Paul Delouvrier, Delegate General in Algeria of the new French Fifth Republic. These new villages were named the "1000 villages of Delouvrier."

Delouvrier launched the process at the time by displaying the positive side of colonisation: "The new village should be a decisive factor in human progress."[7] The use of the word 'village' in place of 'regroupment camp' marked, on the one hand, General de Gaulle's intended shift from military power to civil power, and, on the other hand, the 'humanist-Christian' vision and the sense of duty of the senior official, Paul Delouvrier and his cabinet.

Colonial space is from the beginning organized around the juridification of the master—slave relationship. As indicated by Achille Mbembe: "Recent studies clearly demonstrates that this policy of absolute difference and the logic of segregation that it implies, from beginning to end, has been based on 'dispositifs' for the becoming animal and becoming beast of the other."[8] It is this creation of a colonial subject that allows his maximum exploitation as taxpayer and worker.

For instance, the French government conserved the 'Arab taxes' after the conquest of Algeria. That is to say, these were the taxes settled by the Turkish according to religious law or customs. However, the army 'carelessly destroyed' the land register and the land ownerships registries maintained by this same Turkish authority. Mark Neocleous's observation on the relation between the rise of the high taxation regime and the creation of an indigenous working force confirms the nature of pacification.[9] The use of colonial taxation permitted at the same time the fabrication of wage labour and monetized the colonial society in a simple way.

Transfers of population inside the mainland began with French colonization in 1830 and remained important during the nineteenth century. After any riot or rebellion, villages were destroyed, indigenous populations were forced off their lands, and leaders were imprisoned in penal colonies outside North Africa, mainly in New Caledonia[10] or French Guyana.[11]

Behind the military issues lay the question of territorial planning. The transformation of the rural milieu happened, violently, rapidly, and irrevocably. With the implementation of the "Plan Constantine" and the "Programme of 1000 Villages," settling people through the lens of urban and rural planning became the principal tool for managing the "events,"[12] the euphemism used by the authorities at the time of the Algerian war, closely tied to military action.

Employment and unemployment, union activism and strikes, social activities and sports: all human life was placed under surveillance. Commenting on his own role in the pacification of Algeria, David Galula reports the following exchange:

> In February 1957 (I am not sure of the date), the FLN ordered another general strike all over Algeria. I think it was to coincide with the UN debates on the Algerian question … That evening I received a phone call from Colonel Boissot, the Zone Chief of Staff:
>
> "Congratulations, Galula."
>
> "Thank you Sir, but what for?" Maybe I was being promoted.
>
> "We had a plane flying all over Kabylia today to check on the strike. The only place where the pilot reported any civilian activity was in your *sous-quartier*."[13]

There's a strange side to this discussion that is more in keeping with the media coverage of a social event rather than a discussion about an act of war. This is exactly where we find the benefits of these techniques of counter-insurgency: to establish management over the long term, not only during the period of an acute crisis. This is what Galula theorised by taking as an example the Indochina War: "The victory isn't the destruction of the insurgent forces and their political system in a given region. If the first ones are destroyed, they will be recreated locally by the second; if the two elements are destroyed, they can both be recreated by the action of insurgents from outside. This was the case after each of the cleansing operations carried out by France in the Plaine des Joncs during the Indochina War."

This is the approach that brought architecture and urbanism together as a preventive war operation: to manage territorial planning as a battlefield, as a form of pacification. This was a long-term project. The government in power had to convince the public that "the political and social status quo it embodies corresponded with the aspirations of the population. From this truism followed the principal objective to be set for the loyalist propaganda: to prove that the loyalist status quo was worth more to the people than the revolution preached by the insurgents. The loyalists had therefore to simply invent a counter-cause."[14] There, also, modern architecture had its place provided by water, electricity, or television. The army had been responsible for employment, in substituting working on Public Works Buildings (BTP) for traditional labour in the fields or unemployment, another form of modern social life for those abandoned by the system.

Modernisation, the social advancement principally put in place by territorial planning in the Algerian countryside, was given as the reason that would mobilise the entire nation. We see a concrete implementation of an operation based on the establishment of a contra-cause with the creation of the Thousand Villages but also with the urban organisation of the Constantine Plan, announced by General de Gaulle in person at Constantine on 9 October 1958. We see emerging, in fact, the use of rural and urban planning of the territory, as a tool of managing the population, and most particularly of its utilisation to neutralise part of the population. The construction of the architecture demanded by the military also had to call into question the key phrase of the counter-insurgency: "empty the water to kill the fish." It had to be capable of building an anti-model that could resist insurrection.

The old structure was cleared with the destruction of villages, the loss of livestock, and the impossibility of working lands situated too far away or in forbidden zones. Officers that supported these actions such as Captain Espeisse reported their military efficiency in these terms: "In June 1958 the RCIM and the SAS of Abdelys and of Beni-Quazane created eight identical villages with farms in self-defence. The small Muslim centres saw the light of day by concentrating the residents. Then the countryside were emptied helter-skelter, from housing units constructed around water sources."[15] The idea that modernisation is under way was also a recurring theme: "The new village has resources equal to the old, which are likely to increase; it meets many criteria of progress—a modern habitat, an administrative centre, a participating population, a school, a community centre (sport facility, infirmary)."[16]

Michel Rocard's report is much more alarming and shows the hard and inhuman conditions, with many dramatic cases. This report written in 1958 for Paul Delouvrier, after the visit of about fifteen regroupment camps is worth quoting at length:

> In almost all cases, regroupment camps are gathering a population initially with no earned income. When farm income declines significantly, the military has two options to feed the population: assistance and unemployment worksites. The number of destitute regrouped persons—that is to say here with no resource—is estimated at 200,000. Food intakes distributed as assistance are very small: in one of the cases observed, it was limited to 11 kg of barley per adult and per month, which is very few when there are young children. The biggest problem is the complete absence of regularity of this aid. Non official, due to a civil servant's or an officer's good will, this aid is sometimes interrupted because of the initiator's departure."[17]

In many cases where there were no job prospects or land to cultivate, the army employed men of the village to construct the infrastructures, roads and 'public' buildings. In other camps, food aid was necessary to support daily needs. Many of the villagers were transformed into workers at the service of the State or into unemployed who received food assistance. The conditions were far from those described in the propaganda leaflet distributed in April 1960 and repeated by General Faivre in his book:

> Six hundred villages are under construction or transformation. This policy:
> - extends the municipal reform and puts an end to isolation;
> - accelerates human and social development advancement (education, AMG, social standing, the birth of new elites;
> - accelerates economic development (agricultural cooperation, relations with technicians, making rural life attractive). The aim is to improve performance, educate farmers, and appeal to private companies;
>
> The basic level of this policy is the district, in liaison with the military sector. The choice of location is paramount; it must be near a water source or sink; its legal status must be cleared. The facilities include an administrative and commercial centre, suitable and diverse habitats, and satellite hamlets. The public is concerned; it's necessary to obtain its membership and participation.[18]

This kind of text demonstrates the way in which an entire imaginary construct appears, quasi pious, functioning as a positivist apology for contemporary technologies. The 'new villages' are then presented as 'viable sociological units,' symbolising the 'progress of the Bled' and asserting the French state's intent to modernise Algeria.

The 'village' being created is neither metaphor nor marginal phenomenon. The actual number of displaced persons placed in regroupment camps or in a program like the '1,000 Villages,' calculated at the end of the war in 1962, appears to comprise between 2.1 and 2.3 million, from a total population estimated at 9 million in 1961-1962. The demographer Kamel Kateb estimates 600,000 displaced people living voluntarily in the slums closer to the urban centres to escape the regroupment camps; these two figures give a total of 2.7 to 2.9 million people, equivalent to a third of the total population. Pacification, as Mark Neocleous declares, is here a long-term police operation. It's the administration of the whole territory: watching and securing the territory at all scales and its population in relation to a liberal project.

This architecture casts anew the shadow of Mao Tse-tung's military strategy of cooperation with the 'potential allies' who were the colonial or semi-colonial countries. Mao Tse-tung, map designer: He redrew the world his own way. In 1951, he explained to the whole world China's role in the new wars of decolonisation when he presented on his map of the world two kinds of countries (outside the Communist bloc): The colonial and semi-colonial countries in one colour, and the capitalist countries in another. David Galula sheds light on this map by comparing it to the Treaty of Torsadilla of 1494, in which the Pope separated Spain's and Portugal's areas of potential sovereignty, giving to Spain western America and to Portugal eastern America, Africa, and Asia. Lung Ting-Yi, at the time head of the Propaganda Department of the Central Committee, said explicitly, "The model of revolution to follow in the capitalist countries is that of the October Revolution. The model of revolution to follow in the colonial and semi-colonial countries is that of the Chinese Revolution, the lessons of which are priceless."[19] The colonial and semi-colonial countries were in fact tied to China whereas the capitalist countries were tied to the Soviet Union.

Of course, Mao presented himself not only as the Head of State of the Republic of China, but also as an important military strategist equal to Carl von Clausewitz. In that context, he was also the war leader of the entire Asian zone, ready to handle the logistics of these colonial wars. Via those different facets, his influence was decisive.

If we're interested in his possible influence on architecture and urbanism, we can refer to this sentence: "Many people think it impossible for guerrillas to exist for long in the enemy's rear. Such a belief reveals lack of comprehension of the relationship that should exist between the people and the troops. The former may be likened to water, the latter to the fish that inhabit it."[20] Ann Marlow tells us that the concept articulated in this sentence was publicised in the United States for the first time "in an American policy article in 1943 by George Uhlmann."[21] This analogy, of the fish and the guerrillero, would become popularised very rapidly in the 1960s as a key phrase in the counter-revolutionary war, theorised by the Americans, the English, and the French. The French strategists responded to Mao with, "You have to drain the water to kill the fish."

Around the image of the fish and the water, what is at stake is more than just the fate of the guerrillero. It's the whole territory, the whole landscape, a territory in movement, in mutation, that becomes the challenge of the war. According to the counter-insurgency military theory, the war is demilitarized to become a war of intelligence, a war that takes the city or the territory, as an essential issue. If the fish and the inhabitant are one, it's the pervasiveness of the environment that shelters them here, the water that put architecture and the processes of visibility at the front of the scene. Once again, Mao won, by 'imposing,' in fact, the massive regroupment on a colonial nation, as a response; it's the insurgent who imposed his tactic on the coloniser.

From the perspective of a long and meticulous management of a large territory, we see how architecture and urbanism were central to the pacification of Algeria. Pacification is part of the military art, but the evidence of the Algerian experience shows how far pacification goes, beyond the 'military' and spreads into the wider society, and one of the ways it does so is by means of architecture. Pacification thus become a fundamental modality of urbanism.

The urban environment has to be involved because control of the city is control of the crowd, and to secure the crowds for its cause is regarded as essential for winning the revolutionary war. The experience of Algeria shows the extent to which the doctrine of winning the revolutionary war requires combined control of the population and control of information, that is, control of the urban environment as a priority.

THE DECONSTRUCTION OF THE INDIGENOUS LANDSCAPE

Destruction of the indigenous landscape and construction of a new landscape is a key point in pacification technique. One definition of landscape is given by Fernand Braudel: "Landscapes, spaces are not only present realities, but also, and mainly relics of the past. Earth is, like our skin, doomed to keep the traces of ancient wounds."[22] The stake is precisely there: to erase the past, to erase the conquered population's story and move towards another story. Return to the large landscape helps us realize the strength of the colonial project. The use of large landscape as a tool for pacification goes well beyond the creation of infrastructures or the settlement of intensive farming production units. Landscape becomes the key tool for the aesthetic expression of domination, an aesthetic that uses a double psychological power: a power of conviction by showing off forces and a power of acculturation.

To go further in the understanding of the psychological role of landscape for pacification, this chapter uses a series of images. I spent one week at the Établissement de Communication et de Production Audiovisuelle de la Défense (EPCAD). EPCAD is the French army service that keeps the audiovisual archives of the French Army, supplying the media for all the images produced by the army about contemporary and past conflicts.

Those archives include films and photos. Besides the official archives, many private archives were added. Those photos, based on the research, offer a large vision of the rural and agricultural territory, a territory which is rarely seen on pictures except in colonial agricultural magazines. I took a particular interest in the photos taken by the French Army between 1956 and 1958. They document harvesting in Algeria. Harvesting is done by the French Army itself, protected by elite soldiers and armored units on behalf of big colonists landowners. These images clearly depict the yield obtained from the land, and the exploitation of agriculture for the benefit of the coloniser. But these images also show the critical aspects of the transformation of the landscape. On the one hand, we note the radical planning of this territory in terms of the intense cultivation of cereals. The intensive monoculture is organized to produce the maximum possible yield. On the other hand, a process of transformation of the territory is under way. Aside from the word "Algeria" written on the grain sacks, it seems that we are witnessing a French cereal farming region. Here landscape is clearly a tool to assimilate and acculturate, a tool to deterritorialize its inhabitants.

At that point, architecture is a useful tool to investigate the implementation of spatial devices on large scale. The consideration of several historical pictures, and in particular those from the French Army archives, allows us to have a better idea of the spatial organization of the unbuilt landscape.

The large territory allows an understanding of main landscape entities and ecosystems. Fields, roads, trees, ditches, land limits, all those elements contribute to identify a landscape. But the question is: beyond infrastructure, are non-built elements or micro-architectures organising the landscape by the implementation of small scale spatial 'dispositifs'? Those small scale spatial 'dispositifs' may be:

- non-built signs, such as trees identifying a crossroad or a specific place of interest in the landscape (spring, gathering points), simple stones, big white stones set up as heaps indicating the pathway of the nomadic camps.
- writing, such as roadside billboards and words written in Latin script and in French.

- or micro architecture: Votiv kiosks, micro-scale chapels, carved stones (monolithic cross), crosses or even real architecture like little storage buildings with a pitched roof in red tiles.

Archive Bled 58-461 R20 – June 1958

On these photos, these signs are identifying the territory as belonging to a Christian European community. All these are revealing an absence: the absence of landmarks such as minarets and mosques, the absence of Arabic script, the absence of specific buildings or villages that have been razed. This absence superimposes itself with the presence of new elements. The combined destruction and re-construction of pacification is here at stake as one unique action to build a new environment.

It is necessary to recall the different modes of indigenous land confiscation after the conquest of Algeria in 1830. A great part of the land distributed by the

French State to the colonists came from the confiscation of Muslim religious land called Habous properties, of community land called Arch lands, and of public domain land confiscation called Beylic lands. The other lands are bought at low price by the military and administrative services to indigenous people. Tribes cantonments were an important contribution to the 'civilizing process' put in place. As uncivilized tribes were targeted as a priority social body to be transformed. This new method replaces the simplistic method of the very beginning of colonization that consisted in repulsing tribes and decimating them to monopolize their lands. This first method had the disadvantage to create insecurity. The new method often combines cantonment and the settlement of villages. Restricting tribesmen vital space meant eradicating their existence by making breeding impossible. The cantonment organized by the colonial administration ('Bureau Arabe'), is based on the principle of forest cantonment created during the French Revolution (1790).[23] Forced sells of the surplus of land (identified by the army itself) induce arbitrary transformations that destroy nomad societies' possibility of existence and impose a sedentary capitalist society on the territory.

Archive ALG 56-124 R51 – June 1956

Beyond land dispossession, the nationalization of vacant properties, Habous properties, is breaking the link between communities and their places of worship and their common spaces. Habous properties' confiscation put an end to the funding of mosques and Koranic schools. The school system based on religion soon collapsed. The number of mosques soon reduced: the unused buildings were considered as vacant properties and destroyed. The education system rapidly broke down.

In cities such as Alger, for instance, classrooms were requisitioned to host the French Army. This is concomitant with the transformation of mosques into Catholic churches or with their destruction and replacement with public spaces.[24] Thus the places of worship and teaching disappeared. What was at stake was the destruction of Islam as a source of resistance and counter-power.

Putting the shaping of the landscape into the context of struggles means questioning the link between the economic aim of capitalist exploitation and the colonizer's will of power and domination. The disappearance of indigenous land is never put forward as a plundering but rather as a "redemption," giving the process a Christian character. Culture is also used as a "soft power," including by writers such as Guy de Maupassant: "The Arab's furrow isn't like this beautiful straight and deep furrow of the European ploughman."[25] Social Darwinism is everywhere in texts such as these, used time and again to justify the dispossession of the Arabs. The civilizing mission is precisely to prevent indigenous laziness from spoiling the ground: "This nonchalant farmer would never stop or bend to uproot the parasitic plant that is just in front of him."[26] For this to be done, the indigenous farmer must be replaced: "In our country, the fierce farmer, jealous of his land more than of his wife, pickaxe in hand, would pounce on the enemy grown on his land and, with large woodcutter gestures, would hit the persistent root driven well in the ground, without rest, until it is beaten."[27]

As a consequence, 'progress' justified the strong modifications of the landscape, progress of agriculture itself regarding the selection of cultures but also regarding tools. 'Modern' agricultural plantation is the space of best and more: improvement of yields, better productivity. The porosity between agriculture language and economy language is fascinating: Infrastructure progresses, construction of new roads, new railroads, and also hydraulic infrastructures, medical and social progress with the construction of hospitals, maternity hospitals and vaccination services—the whole of this progress is brought forward by the European civilizing mission. But it can also be seen as support structures allowing the territorial economic exploitation, in particular for vineyards and grain lands. The development of thousands of hectares in the Plaine de la Mina for instance is indissociable from the hydraulic works that allowed irrigation.

Archive Bled 56-83 33 - June 1956

Archive Extract Bled 58 461

One Archive Album being examined in ECPAD

CONCLUSION

The role of capitalism is to model and implement accumulation cycles. In this model, the place of the territory is separate. This space is in fact gathering important risk factors: population, natural hazards, climatic data, geostrategic risks. We know that risks related to a population's governability are the most important. Those risks as well as the geostrategic ones can be lowered thanks to a work of prevention, whereas natural hazards and climate risks are completely unforeseeable.

Programming the landscape can then be understood as a strong psychological work that uses landscape as a tool for population's management. The relation between capitalism and landscape has been rarely considered and analysed as such. Landscape has often been seen as a vector of popular cohesion allowing the implementation of a strong nationalism. From 1933, Nazi Germany carried this argument further, working on a Germanization of its territory's landscape, and then from Poland's invasion working on landscape's Germanization on the whole Polish occupied territory. Population's segregation in urban, peri-urban, or rural areas is another vector that has been broadly documented in the USA and in South Africa.

Beyond those vectors, programming the landscape concerns the production or reproduction of labour, and the political and social conditioning of a population and the conditions of its governability. Long-term labour management has always been crucial. We can observe a set of procedures of the landscape transformation named programming that aims to the operationalization of large scale violent processes of acculturation, landmark modifications, changing signs, all to deterritorialize the inhabitant from his or her own living environment. Those new procedures are juxtaposed to more classical ones like relocalisation and population transfer, tax system implementation, and the creation of signs linked to the dominating culture.

It is this whole environment linked to the material and immaterial culture of a population that territory programming is destroying. If writers such as Toni Morrison or Edouard Glissant, for instance, succeeded in giving voice to landscape, it's to make this lifeless voice of the dominated, of the dominated cultures ruined by capitalism, heard. Through a landscape filter, we understand how those long, slow, and totalizing processes operate at large scale to reduce potential resistances and annihilate differences.[28]

NOTES

1. James Scott, *The Art of Not Being Governed* (New Haven and London: Yale University Press, 2009), 150.
2. Fabien Sacriste, "Surveiller et Moderniser: Les Camps de 'Regroupement' de Ruraux Pendant la Guerre d'Indépendance Algérienne," *Métropolitiques*, February 15, 2012, 2, accessed August, 2012, http://www.metropolitiques.eu/Surveiller-et-moderniser-Les-camps.html.
3. See Stuart Schrader and Ryan Toews, in this volume.
4. Michel Cornaton, *Les Camps de Regroupement de la Guerre d'Algérie* (Paris: Editons l'Harmattan, 1998), 109.
5. Gilles Deleuze and Felix Guattari, *Mille Plateaux* (Paris: Editions de Minuit, 1980), 479.
6. This work of managing indigenous people intensified after World War II. In 1947, during the repression of the riots in Madagascar, the French government tested forced relocations, along with new techniques of psychological warfare, such as throwing prisoners taken at random out of airplanes in flight to terrify local populations. Likewise in Cambodia and Indochina the French army experimented the regroupment of populations.
7. General Maurice Faivre, *The 1000 Villages of Delouvrier, Protection of the Muslim Populations Against the FLN* (Sceaux: Éditions l'Esprit due Livre, 2009), 51.
8. Achille Mbembe, "La République et l'impensé de la « race »," in *La Fracture Coloniale: La Société Française au Prisme de l'héritage Colonial*, eds. Pascal Blanchard, Nicolas Bancel and Sandrine Lemaire (Paris: Editions de la Découverte, 2005).
9. Mark Neocleous, *War Power, Police Power* (Edinburgh: Edinburgh University Press, 2014), 145-149.
10. Caledonia was a very politicised penal colony dedicated to French revolutionary repression during the XIXth century. 2166 Algerian people were concerned by transportation or deportation (ordinary prisoner and political prisoner) in this penal colony.
11. Guyana penal colonies was focused for a long period for prisoners coming from the colonial space. The estimates give a total number of prisoners equivalent to 17,000 people in this penal colony from which 25% were Algerian indigenous subjects.
12. The euphemism used by the authorities at the time of the Algerian war to reduce it to a police action against terrorists.
13. David Galula, *Pacification in Algeria, 1956-1958* (Santa Monica, CA: RAND Corporation, 2006) 135-136.
14. David Galula, *Counter-Insurrection: Theory and Practise* (Paris: Editions Economica, 2008), 115.
15. For example, General Maurice Faivre, *Les 1000 Villages de Delouvrier, Protection des Populations Musulmanes Contre le FLN*, 74.
16. For example, General Maurice Faivre, *Les 1000 Villages de Delouvrier, Protection des Populations Musulmanes Contre le FLN*, 51.
17. Michel Rocard, *Rapport sur les Camps de Regroupement et Autres Textes sur la Guerre d'Algérie* (Paris: Editions Mille et Une Nuits, 2003), 132.
18. Faivre, *Les 1000 Villages de Delouvrier*, 61.
19. David Galula, *Counter-Insurrection Theory and Practise* (Paris: Éditions Economica, 2008), 206.
20. Mao Tse-tung, *On Guerrilla Warfare*, chap. 6 (Selected Works of Mao Tse-tung: Vol. IX, 1937), http://www.marxists.org/reference/archive/mao/works/1937/guerrilla-warfare/ch06.htm.
21. Ann Marlowe and David Galula, His Life and Intellectual Context (Carlisle: Strategic Studies Institute, 2010) [28]
22. Fernand Braudel, *L'Identité de la France* (Paris: Editions Arthaud, 1986), 25.
23. The Cantonment in favour of the municipalities consisted in exchanging a right of use on the whole forest for a property on a part of this one and giving up the right of use of the whole. The land created is bounded. The application of this law destroys the right of use. The creation of the property right is clearly opposed to the right of use.
24. Government square in Alger was built in 1832 by the military engineering services at the place of As-Sayyida mosque.
25. Guy de Maupassant, *La Vie Errante*, (Paris: Edition Paul Ollendorff, 1890), 177.
26. Guy de Maupassant, *La Vie Errante*, 177.
27. Guy de Maupassant, *La Vie Errante*, 178.

28 I thank Tyler Wall and Mark Neocleous for their works organizing the preliminary seminar that made this book possible. My gratitude goes to the French army archive centre—Etablissement de Communication et de Production Audiovisuelle de la Défense in Paris—for facilitating this research.

Urban Pacification and "Blitzes" in Contemporary Johannesburg

Christopher McMichael

The term pacification regularly appears in scholarship on securitization and militarization in cities. Stephen Graham refers to both an historical process in which the suppression of class conflict in European cities in the nineteenth century drew extensively on 'explicitly colonial models of pacification' and warfare honed in the global South and contemporary experiments with facial recognition software in British cities, which parallel experiments to 'pacify urban insurgencies in Iraq.'[1] Pacification is often used when discussing the rehabilitation of counterinsurgency doctrine in the 'Long War,' while the 'rhetoric of military pacification with respect to the chaotic spaces of Iraq and Afghanistan' has been compared to 'zero tolerance' policing in US cities.[2] David Harvey uses the term to suggest more subtle forms of embourgeoisment with political pacification through consumption and 'pacification by cappuccino.'[3] But while this term is ubiquitous, it is not defined or developed as a concept in its own right, and is rather casually used as a euphemism for social control, which has sinister connotations.

This chapter addresses this gap and argues that it should be a key concept for understanding how state power and capitalism are reproduced in cities. This concept will then be applied to a study of Johannesburg, where the securing of 'chaotic spaces' is an abiding fixation of state rhetoric and practice. Of course this is not a fixation unique to Johannesburg, but it is a city where anxieties about violent crime and policing have special prominence. One of the most prevalent security tactics in Johannesburg is the police-lead blitz, which this article will argue is *a pacifying tactic* of forcefully and often violently regulating space.

The recent critical reappropriation of the term pacification emerges from the anti-security perspective, which argues that rather than signifying a politically neutral human need for protection, security is an intrinsically authoritarian concept used by the state and capital to extend domination, manage unequal social orders, and to disguise the systemic causes of 'insecurity.'[4] The concept of security is so hopelessly complicit with power, as evidenced all too well by the 'war on terror,' that rather than trying to outline more 'human' or 'just' forms of security, critical scholars need to find an alternative language that will not become complicit with the exercise of state power.

Before doing this, however, we need a better understanding of the vision of security held by the state and capital. Neocleous has called for a reassessment of the term pacification, which historically is associated with European colonialism, going back to the Spanish conquest in the sixteenth century, and counterinsurgency warfare, particularly the American war on Vietnam and the French occupation of Algeria.[5] Returning to historical texts of pacification shows it combined both destructive and productive aspects. It was destructive in the use of police and military violence to destroy or supress resistance but also productive in building a social order of docile populations and controlled spaces.[6] The violence of pacification was therefore understood as a constructive, civilizing violence "war, dressed up as peace."[7] Moreover, both violence and efforts to win compliance are combined with the specific aim of fabricating *capitalist order*, from creating the conditions for accumulation to sustaining accompanying class, racial, gender, and other forms of domination. But this is open-ended and *never finished*, as authorities come against new "disorders" and resistance. While "security" is worked out differently within different cities and states, the concept of pacification highlights how its overarching logic connects different institutions (police and military forces, private security companies and mercenary firms, NGOs, and the administrative arms of the state) and the "domestic" and "foreign." For example, the contemporary wars on drugs and terror link military actions and invasions with internal policing and the "growth of exclusion zones, social fortressing and the militarization of urban spaces" and "guerrilla war against the poor and any categories of population thought problematic to bourgeois order."[8] Urban spaces have always proved a particular site of strategic interest and challenge, as they are both dense concentrations of wealth and power but also of "dangerous populations," transgression, and revolt.

Returning to urban scholarship, it appears that these themes are very much implicit within the literature, which mentions pacification, although they have yet to be explicitly drawn out. Examples taken from a variety of different contexts and eras highlight this thread of pacification in the name of order. In early twentieth century Barcelona the ruling classes were involved in a bitter conflict with the Anarchist counter-culture of the city's working class: "bourgeois consciousness became predicated upon a dread of urban disorder and the desire to pacify and reconquer a city besieged by an army of proletarian barbarians."[9] Measures taken included brutal policing, heightened surveillance of the public sphere, and moralizing rhetoric about the social contamination of dangerous groups and its threat to industry and the work ethic.

In the 1960s, "American experiments in domestic pacification" drew upon a counterinsurgency efforts in South Vietnam and elsewhere, predicated on fears about growing militancy in black urban areas, with "Saigon and Harlem" seen as two fronts in the same conflict.¹⁰ This inspired a variety of poverty programs in urban areas intended to pre-empt domestic revolt and communist insurgency by winning loyalty to the capitalist order, thus drawing on colonial practices that regarded state force as creating the conditions for the "truest sign of pacification— the peaceful animation of roads and markets."¹¹ The pacification of cities later expanded to include the violent FBI and police repression of militant urban movements. Later, the language of Vietnam was weaved through the "wars" on drugs and gangs, with the "pacification of Los Angeles" predicated on a war against gangs regarded as a local "Viet Cong."¹² At the tactical level, this saw the forces of the LAPD marshalled into publicity-drawing sweeps "like the Marines hitting the beach at the beginning of LBJ's escalation in Vietnam, the first of the thousand cop blitzkriegs made the war in South Central LA look deceptively easy."¹³

More recently, the state government in Rio De Janeiro has embarked on a Pacifiying Police Units (UPPs) program, which is officially described as the state seizing back controls of *favelas* from drug gangs so that services and order can be established. However, critics have argued that the program is actually about attracting capital and investment in the wake of the World Cup and Olympics, and it has been marred by evidence of endemic police violence against favela inhabitants.¹⁴ Notably, the Brazilian program was compared to US counterinsurgency operations in Iraq and Afghanistan in a leaked diplomatic cable.¹⁵ Informal spaces in cities from "Iraq to the favelas of Rio" are viewed as challenges to social order, with the prescription being a combination of disciplinary actions and urban development.¹⁶ Such a carrot and stick approach is especially pacifying, as demonstrated by left-wing governments and NGOs in Latin America, which have used the expansion of services as a "mechanism which strengths power and domination" by containing radical aspirations for systemic change.¹⁷ Finally, the term pacification has also been used in reference to the various antagonistic movements organised against austerity and police violence. One collective reflection on Occupy Oakland suggests an historical transmutation where the war on drugs against the

> Unruly population, especially poor people of colour in urban areas ... has culminated in a 21st century imperial war on terror. The equipment and tactics of urban pacification are now being turned on American cities and on the citizens and non-citizens who are targeted by austerity methods which have for decades been applied to the Global South.¹⁸

This article will therefore contend that this idea of urban pacification should be a central category in understanding the pernicious grip of security over city life. Certainly, a wide body of literature already discusses how different combinations of coercion and consent building are combined to build urban order. These include a focus on the biopolitical aspects of urban policing and the project of creating docile spaces and subjects, "military urbanism" in which new surveillance and control technologies both protect the circuit of capital and are products of a lucrative global security industry, and the revival of counterinsurgency in which the winning of popular support for state operations is envisioned as crucial to systemic stability.[19] While the concept of pacification draws substantial inspiration from this body of theory, what makes it distinct is a far more radical critique of state, capital, and the logic of security.

Much recent scholarship has claimed that historical or traditional demarcations between war and policing have been blurred as evidenced by the militarization of urban spaces and of police. In contrast, pacification holds that this boundary is in fact a myth. Historically, violence and production of space in the "colony" and "metropole" served the same ends of class domination, in a unified project of imperial war and policing in pursuit of accumulation. Tactics of war and terror used in colonial cities were adapted to police the working class in Europe—and in times of intense social conflict to crush resistance. Rather than a "boomerang effect," in which tactics used in the South reappear to ensure social control in cities of the North, we can think of a continuum of pacification, underpinned by the dream of ensuring cities of docile proletarians and occupied peoples, governed spaces of work and production.[20] And because capital accumulation is an open-ended process in which all aspects of city life are colonised by the market, this continuum of power is still with us today. By rejecting the epistemological distinction of an absolute separation between war and policing, pacification also rejects the idea that we should look for more "humane" or "civilized" forms of security, or try to replace "militarized" police with more friendly "community"-orientated forces. Instead it holds that hard and soft power are intertwined. While the threat of state violence is always present, strategies of governance are contoured to different spaces, so that while overt repression may be favoured in one, consensual measures are adopted in another. An analysis of pacification therefore proposes to untangle the threads and discover why some measures work in some places better than in others.

This article will therefore untangle the threads in the context of Johannesburg, South Africa's largest city and the biggest metropolitan economy in Sub-Saharan Africa. My key focus is on using the concept of pacification to uncover the political and material prerogatives of security policy and practices in the city. During 2013, I began post-doctoral research on public safety policy in the city and noticed that the concept of "blitzes" was consistently referenced in official documents and reports. I was intrigued due to its historical origins in air war, and decided to focus my research on the usage and meaning of the term. In other words, is the fact that a term derived from warfare so ubiquitous in the everyday civil policing in Johannesburg just coincidental, or does this ubiquity highlight the continuum of war, police, and accumulation that is pacification? Between 2013 and 2014, I conducted a series of semi-structured interviews with current and former city officials, who I approached as they had key roles in the policy and operational aspects of public safety in Johannesburg, which focused on security generally and "blitzes" more specifically. All the respondents agreed at the beginning of interviews that this information could be used for academic purposes. I also collected a chronological archive of media articles, press releases and local government reports that mentioned the term "blitzes" in Johannesburg. This methodology focuses on the ideological and discursive construction of security. Therefore, this article does not look at the ethnographic or tactical intricacies of policing in Johannesburg, nor, on the other hand, on the experiences of people who have been the target of blitzes.

Along with advancing the concept of urban pacification more generally, this article also has a more specific polemical intent. In post-Apartheid South Africa, security—broadly and vaguely defined as protection from crime and violence—enjoys hegemonic status in government, the media, and academia. But in practice, security is the exact pretext the state uses to justify increasingly authoritarian and violent attacks on predominantly the black poor and working class, most disturbingly evidenced in the 2012 Marikana massacre where striking miners were gunned down by the SAPS (South African Police Service). So rather than attempting to suggest better forms of security than blitzes, this article offers a counter-hegemonic narrative that traces the pacifying imperatives behind public safety and policing.

BLITZES

In October 2013, the City of Johannesburg (COJ) launched a "Mayoral Clean Sweep Initiative" with a series of co-ordinated police raids throughout the city. They were conducted by the Johannesburg Metropolitan Police Department (JMPD), assisted by various government departments, and intended as the opening salvo of a campaign aimed against "bad buildings, land invaders and shack dwellers," extending to reconfigurations of the built environment with the rapid demolition of a multi-story taxi rank in Bree Street.[21] "Clean Sweep" was widely condemned by both street traders, unions and academics as economic warfare against the working poor. Despite the city's claim that the clampdown was aimed at illegality, traders with formal permits were indiscriminately arrested alongside those without, in violent scenes which included the JMPD using whips and dragging people out of vehicles.[22] According to some figures, 6000 traders were evicted causing a loss of livelihood that affected, in turn, thousands of the hawker's dependents.[23] The struggle against the removals was conducted through court interdicts and street protests. Two months after the initial raids, the national Constitutional Court ruled that the City's actions were illegal and that registered traders could return to their stalls. Although "Clean Sweep" was suspended, the

JPMD used rubber bullets to disperse returning traders and arrested a lawyer representing them.[24]

The conflict between street traders and the city government is ongoing, and the events around Operation Clean Sweep reveal important facets of the governance of Johannesburg. Firstly, it showed the power of the state to rapidly and dramatically deploy its forces and materiel in urban spaces, from police interventions enforced with whips to the destruction of buildings. Secondly, these actions were explained with reference not simply to law enforcement but to the general order and health of Johannesburg, in which all kinds of illegality and unregulated activity are depicted as potential threats to the maintenance of a stable urban geography. For example, one COJ council document lists "the challenges" presented by informal trading as including the hindering of police work by congestion, obstruction to CCTV cameras, and increased litter.[25] However, it also highlighted the constraints on state power as social protest forced the authorities to formally suspend Clean Sweep. Finally, the police repression of informal trade resonates with urban struggles elsewhere—such repression lead to the self-immolation of street trader Mohamed Bouazizi that catalysed the Arab Spring.

As suggested by Philip Harrison, former executive director of urban management for the COJ, Clean Sweep was viewed by officials as a response to the perceived failure of prior campaigns to enforce order in the city, reflected in its unprecedented city-wide scope.[26] In the past the COJ has relied on a more limited set of "occasional blitzes, campaigns and joint operations" targeted at problem areas.[27] The term blitz, in particular, has a powerful association with force, power, surprise, and speed and (from the German word for lightning) appears to have entered into the English language in World War Two, as journalists popularized the term "blitzkrieg" (lightning war) to describe the devastating air and land tactics used by Nazi Germany to conquer Europe. Although many military historians argue that Blitzkrieg was never an official doctrine of the Wermacht, it has persisted as a term to describe the combined usage of aircraft, armour, and infantry for rapid assault.[28] With the air raids on England in 1940, the abbreviation "blitz" therefore came to signal "an intensive or sudden military attack" (Oxford English Dictionary). The etymology of the word situates it within the repertoire of aerial state terror "from the screeching of bombers during World War Two" to the Shock and Awe strikes used during the 2003 invasion of Iraq, intended to create a feeling of fear and unease amongst civilian populations.[29] More informally, it describes a "sudden concerted effort to deal with something"—such as this example given in the OED: "Katrina and I had a blitz on the cleaning."

Despite this more informal usage, the term has maintained a strong association with police and military power. In a 1979 instalment of Judge Dredd, a long running action-satire comic of police authoritarianism (parts of the 2012 film adaptation were shot in Johannesburg), random citizens are selected for "the crime blitz." One target attempts to hide his misdeeds with the ingratiating "I understand,

Judge Dredd, the crime blitz hurts only the guilty. It is very useful for uncovering secret illegal activities and for discouraging crime." More recently, we can read of both "traffic blitzes" aimed at catching drunk drivers in Western Australia, and a joint US-Italian "anti-Mafia blitz" in New York.[30] In 2013, a South African comedy film called *Blitz Patrollie* followed the adventures of two Johannesburg cops. From the military side, contemporary references include "small scale lightning raids" such as the killing of Bin Laden, favoured by US security agencies and the 2006 "intense public relations blitz" used by the Pentagon to promote its counterinsurgency strategy in Iraq.[31] Within the term blitz we find overlaps between warfare, policing, and general administration ("blitz on cleaning"), all of which pivot around the sudden use of force to advance particular goals, from overwhelming the enemy to cleaning our homes. What the different usages highlight is that the blitz is a type of raid, a surprise incursion. And historically, the raid has been central to pacification, such as raids on horseback in colonial territories, air raids in the twentieth century and the common tactic of the police-raid, which catches targets when they are "least aware."[32] The air raid works in tandem with ground operations, such as in the war on Vietnam where US pacification plans on the ground were accompanied by regular aerial blitzes against Vietnam, Laos, and Cambodia.

Passing mention is made of blitzes in research on urban governance and politics in post-Apartheid Johannesburg. Generally, "anti-crime" blitzes have included mass arrests, roadblocks and contraband seizures in the city.[33] While they are broadly aimed at "non-compliant" subjects, other research has suggested that such measures are targeted at more specific groups.[34] Blitzes have been discussed within the context of the "scapegoating and criminalization" of undocumented migrants from other African countries with police crackdowns on immigrant businesses intended to display government's "tough" stance on "illegals."[35] For Martin J. Murray, blitzes are a tactic deployed more generally against the poor in Johannesburg, which serve both practical and theatrical purposes:

> Law enforcement agencies have operated like an occupying army and viewed inner-city residential neighbourhoods as a domestic warzone, dangerous places that harbour an alien population stripped of customary legal protections and privileges. Routine crime blitzes function as staged performances enacted to foster the dual conceit that, first, the police have effectively removed criminals from the streets, and, second that the city is a safer place as a consequence.[36]

While this dystopian phrasing in the excerpt above paints a vivid picture of the brutal social inequalities characterizing daily life for many in South Africa, its primary focus is on the coercive and exclusionary exercise of state power. However, the blitz terminology is used more broadly to refer to state actions that appear to be in an entirely different and more benign category than punitive class policing. This includes actions such as a "mega-blitz" to refurbish and clean up parks in Soweto after heavy rain fall and a blitz aimed at "encouraging" factory owners to adhere to environmental and health regulations.[37] Furthermore blitz operations are not just aimed at the defenceless urban poor, with others being targeted against well-armed and organised crime syndicates to "pacify … urban centres, where violence had spiralled out of control."[38]

INTENSIFIED OPERATIONS

According to Chief Superintendent Wayne Minnaar, spokesperson for the JMPD, the police blitz in Johannesburg has "always been used, and always will be used when there is a need for it."[39] This need arises when the non-compliance with the law "becomes out of control" necessitating "intensified" police intervention—this includes such scenarios as hawkers not complying with regulations and "illegal land invasions" by squatters. But while, as I will shortly argue, the term does seem to have a more recent origin than that presented by Minnaar, throughout Johannesburg's history the city's rulers have indeed relied on "intensified" operations to control "non-compliance" and moral and social "decay." For much of its existence, Johannesburg has stood as a symbol of the state violence and racial social engineering that has characterised South African history.

After the city's founding in 1886, Boer and later British authorities facilitated the growth of industrial capitalism based on gold mining through intensified social control such as segregation between black and white labour, as well as more generally attempting to separate "labouring classes from the dangerous classes."[40] A regular tactic of enforcement over dangerous spaces and bodies was the lightning-fast deployment of state forces in the name of upholding order and health. For example, after a breakout of pneumonic plague in 1904, the multi-racial "Coolie Location" slum was destroyed with "military precision," while a police report from 1919 notes the "annual raids" launched upon the heavily controlled mining compounds for black labourers.[41] Police and military raids were also used in response to open class warfare, which culminated in the aerial bombardment of white strikers during the 1922 insurrection. In fact, proponents of air power in the 1920s regarded it as a police technology for securing territory and monitoring and suppressing colonial populations and internal "dangerous classes"—a year prior to the air raids in Johannesburg, striking miners in West Virginia were bombed from the air.[42] A major advocate for aerial pacification was South African General Jan Smuts, who was involved in the establishment of the air ministry in Britain during World War One, and who was "aware of the importance of aeroplanes as a means of control and submission."[43] Smuts was prepared to use aeroplanes against striking black miners on the Rand in 1919, and while they were not used in that instance, the deployment in 1922 was followed the next year by bombings against a revolt in Namibia.

Throughout the twentieth century, policing in Johannesburg was characterized by what Irene Mafune, director of Region F (including the Inner City) called the "kick in the door" approach.[44] Surprise assaults were used to clamp down on underground apartheid opposition movements: Nelson Mandela was arrested in the 1963 raid on Lilliesleaf Farm, to name one such incident. The routine police raids were used to administer the system of pass laws and influx control, which drastically restricted and controlled the black majority of South Africans" access to urban space.[45] But as mass resistance to Apartheid increased, the state adopted new "security" strategies. In the wake of the 1976 Soweto uprising, the government implemented counterinsurgency ideas derived from the French and US militaries and attempted to pacify militancy through a combination of selected infrastructural upgrades in townships *and* military campaigns against neighbouring countries and police death squads that hunted down activists. This included the "oil spot" plan of "establishment of strategic bases until the whole area is recovered," a term used in pacification in Algeria and later Vietnam. At the time, this was summarized by a foreign diplomat working in South Africa as a scheme of "crush, create, negotiate."[46]

By the 1980s, the policing system of urban Apartheid began to break down, and in particular, the make-up of Johannesburg's inner city began to change from a white business centre to a residential area for the black poor and working class. In the two decades since the end of apartheid, Johannesburg has become a dense and complex urban environment, with both rapid changes and the continued impact of the socio-spatial legacy of segregation. While it serves as a cosmopolitan space of opportunity for migrants from around the world, it also features a massive gulf between the worlds of rich and poor. Spatially this contrast is most evident in the difference between the corporate headquarters and five star hotels in Sandton, one of the wealthiest areas in Africa, and the neighbouring township of Alexandra which continues to "suffer the ignominy of poverty and overcrowding."[47] Anxieties around violent crime and security have further reconfigured urban geography with assemblages of gated estates and housing complexes, road closures, barbed wired and electrified fences, armed response companies, and city improvement districts, which have even lead officials to worry about the consequences of this "paramilitary approach" to urban space.[48] This dense security apparatus forms the backdrop to everyday life and commerce, police helicopters and private security guards with semi-automatic machine guns in a sprawl of shopping malls, parking lots, and evangelical churches.

A less overt but just as substantial theme in Johannesburg's recent history is changes in how it is formally governed. After 1994, the process began of creating a representative and unified governmental structure out of the fragmented Apartheid system of well-funded white localities and under-resourced black areas.

Beginning in 1996, the COJ was officially amalgamated from 13 different institutions. This amalgamation occurred amidst a sense of crisis and concerns about order, a theme which appears to stay constant in the institutional development of the city in the last two decades.[49] Along with internal conflicts pertaining to these administrative shifts, the COJ was faced with financial bankruptcy due to years of financial mismanagement and rent boycotts from the mass resistance to Apartheid.[50] In particular the inner city became the "geographic symbol" of urban decay for local government and capital due to crime, disrepair, and general urban management issues.[51] Nightmarish images of the inner city even circulated abroad. In a widely read article on "feral cities" that apparently pose a threat to geo-political stability, US Naval college Richard. J. Norton described the inner city as "on the knife edge ... police in Johannesburg are waging a desperate war for control of their city, and it is not clear whether they will win."[52] This exaggerated depiction of Johannesburg as a potential failed city justifies the view of the global South as a series of blighted wastelands in which the US military must be prepared to intervene to maintain world order. And it also omits an interrogation of the way Johannesburg's economy is tied into global circuits of militarism, such as in mercenary companies which send "security contractors," often with backgrounds in Apartheid-era special forces, to Iraq and more recently Somalia.[53]

The newly constituted municipal government of Johannesburg made plans to stabilize its city and prepare the conditions for more sustained economic growth. Asserting state control, particularly in the inner city, was intended both to win legitimacy from the local electorate and to attract external investment.[54] Officials become particularly enamoured with the idea of Johannesburg as a "world-class African city," the Joburg 2030 strategy articulating this vision of a prosperous metropolis. The document describes a "safety and security vision" orientated more around "economic than social goals," based on the premise that reduced crime and by-law infringement would increase investor confidence and thus lead to general improved standards of living.[55] This strategy was directly modelled on the apparent success of the "New York City example ... of no-tolerance" and called for expanded operations not only for serious crimes but for "any infringement of any law at a municipal level."[56] The newly formed JMPD were tasked with using by-law enforcement to establish a law-abiding "mind-set in the population... for the law at every level."[57] As has been extensively documented elsewhere the slogan of "zero tolerance" derived from America was a police slogan which was diffused throughout the world—although its actual success in bringing down crime rates in New York has been contested by many scholars, and in practise it was not really a specific concept but rather an amalgamation of practices that penalised the antisocial behaviour of the lower classes.[58] Zero tolerance was also discursively linked to the conservative "broken windows" theory, which calls for stricter policing of minor incivilities—although this link has been disputed by practitioners who noted that aggressive street policing was not informed by any criminological theory. Regardless, American slogans were seized upon by South African officials and solidified by official exchanges to New York which offered confirmation that Johannesburg needed to adopt such world class policing.[59]

The population and territory which garnered the most focus was the inner city. Operational programs such as the Better Buildings Program and the Inner City Taskforce were established to make the region a more "functional property market."[60] Pacifying the inner city was therefore not just about policing peoples but also using the state apparatus to create safe spaces for accumulation.[61] This regeneration effort was often depicted by its agents as involving the taming of a dangerous urban frontier. The Inner City Task Force called its meeting area "the war room," in which weapons seized during clean-up operations were put on display.[62] These clean-ups were conducted using a "cowboy" approach of brute force, with many of the staffers of projects coming from security forces background.[63] This "militaristic" approach was especially evident in the outsourcing of evictions to the Red Ants, a company specialising in brutal "early morning rousts."[64] Regular raids both provided information on a complex operational environment and asserted state power over unregulated and illicit spaces.[65] But these high-visibility incursions were not exclusively the product of a top-down vision of order. In many cases, they benefited the personal interests of state officials: from functionaries extorting bribes from undocumented immigrants to the more sophisticated collusion between officials, developers, and security companies benefiting from the seizure and resale of dilapidated but valuable buildings.[66]

The reliance on overt coercion led to an apparent split in the city administration between "doves" calling for more focus on housing and service provision and "hawks" fixated on showing zero tolerance to crime and indigence.[67] Recent official strategies prefer a more sophisticated approach, employing the language of creating safe spaces and inclusion over a blunt focus on compliance with the law. For example, the Joburg 2040 document argues that urban crime control has created "islands of exclusion" to the detriment of the "vulnerable" in society.[68] However, this focus on inclusion is also underlain by anxieties about the threat posed to social cohesion by the dangerous masses—in this case the "ring of fire" or the townships and informal settlements around cities which regularly explode into protest and confrontation with the local state.[69] And as seen in Operation Clean Sweep, the rhetoric of inclusion has not impeded the state's reliance on iron fist clampdowns on "problem" spaces. Furthermore, the post-apartheid adoption of tough policing has rehabilitated racist policing under a new veneer, with echoes of the past seen in large groups of "armed police invading recreational establishments at which black men gather: groups of police picking individuals off the streets and throwing them against a wall."[70] Racially charged policing is also intermingled with class bias, as blitzes and other operations were overwhelmingly aimed at urban spaces considered to be black and poor, with similar raids not be launched in more affluent areas—the upshot of that certain crimes would merely be displaced to such less policed areas.[71]

ZERO TOLERANCE

Within this context of the state's demand for order, the term "blitz" has become ubiquitous. It is unclear when the term became a part bureaucratic parlance, or whether it emerged directly from the police or from media descriptions of their activities. Nazira Cachalia, the program manager for the City Safety Program, suggests that the term has been "embedded in the city" and most likely emerges from the Apartheid regime.[72] One exact reference to the term can be found from June 1991, when the government launched a national crime sweep called Operation Blitz in which the police and military searched "8000 homes and 30,000 vehicles."[73] Post-1994, the term emerges in connection with the new government's war on crime, with an essay by Lindsay Bremner, originally published in 1998, specifically listing "police blitzes" as one of the security initiatives prevalent in Johannesburg.[74] The earliest media reference to a state operation as a "blitz" found during my research was from just before the turn of the century, with an article discussing a "massive crime blitz" in Hillbrow ahead of the millennium celebrations.[75] Highlighting the influence of American policing rhetoric, a police official described the operation as a "zero tolerance" initiative that targeted petty offences to impact upon major crimes. This connection is made even more significant because in March 2000 the state launched a "national crime blitz" as part of a countrywide Operation Crackdown.[76] Operation Crackdown was modelled on high profile operations in the US and focused police and military deployments in 140 crime hotspots throughout the country.[77]

The blitzes in March were therefore the first round of this operation and aimed at the neighbourhoods of Hillbrow, Yeoville, and Berea. While this was presented as the state seizing back spaces from dangerous criminal syndicates, in practise anyone in a hot spot could face the force of the "blitz": the government was forced to apologise after detaining both asylum seekers and citizens who were only released after showing proof of identity.[78] Highlighting the term's origins in aerial terror, blitzes also created fear amongst those in targeted areas, with one woman in an informal settlement saying that an early morning raid had her hands trembling: "I was so terrified with all these helicopters flying overhead. Then the banging on the door."[79] Operation Crackdown was also an opportunity to highlight the government's "zero tolerance" stance towards "illegal immigrants" from elsewhere in Africa, reflective of (undeclared) official xenophobia towards migrants.[80]

Such bombast was central to these "zero tolerance" raids as it loudly declared the state's capacity to reclaim the streets in a highly public manner. The dimensions of this capacity are captured in an unusually detailed newspaper article called "Hotel from hell shut down in Hillbrow Blitz" of an operation against the Mimosa

Hotel in November 2001, which according to the state was a threat to security and hygiene ("the place is full of drugs and illegal activities—you can smell it for miles," as well as "decent citizens."[81] The response was a joint police and military "surprise attack from the air and ground" featuring police and soldiers dropping from the roof from helicopters and prying open safes with angle grinders. Government ministers arrived at the scene later on to inspect seized caches of drugs, stolen cars, and forged documents. Although such operations go by different names, this blitz is comparable to anti-crime operations in cities across the world, in which the saturation of spaces with armed units is also conceptualized as a public relations exercise that highlights the proclaimed ruthlessness of the state in hunting down irredeemable miscreants. For instance, this includes what Mike Davis calls "designer drug raids," such as a 1989 SWAT operation by the LAPD which was observed by Nancy Reagan.[82] More generally, this operation highlights the assymetery of power in the blitz, as the full power of the state is used to surprise and disrupt spaces of illegal economic activity and illicit consumption.

LUNCH PACKS AND BULLET PROOF VESTS

The usage of the blitz terminology began to expand. While it was still presented as a rapid state intervention, it began to involve a wider swath of state institutions than just police-military assaults. In March 2001, a waste management blitz was organised to pick up litter that was explicitly conceptualised as a "deployment … Rolled out with military precision, which will leave the city sparkling clean and mark a turning point in the cleanliness."[83] By 2004, the Inner City Regeneration Strategy Business Plan document defined "Blitz operations" as any

> Multi-disciplinary operations (including a range of other law enforcement agencies) in identified areas where severe decay exists including breakdown of law and order which have a severe effect on the whole city e.g. Hillbrow. This includes a dedicated outside enforcement team focusing on specific areas.

Along with a police presence, these operations entailed the inspection of inner city building stock by health and fire officials to serve notice for evictions and by-law infringements. According to the city, such an approach was needed to save buildings that had become "sinkholes" infecting the public environment with crime and degradation.[84] But the "almost military urgency" of health and safety blitzes was also hastened by the economic imperatives of regeneration, with seized buildings being sold off to developers or incorporated into the government housing stock-creating supposedly "low cost" housing which was often practically "well beyond the means of the evicted."[85] The linking of the blitz to both health and law enforcement meant that it came to be viewed as the appropriate response to any situation where the state had to forcefully intervene. A progress report from 2010 prescribes blitz operations for both organized crime and for clean-ups at hostels "to demonstrate commitment to improving living conditions."[86] The latter highlights how the city presented blitzes as evidence of its ability to manage urban space for all citizens, rather than just commercial interests. A 24-hour blitz in 2008 was described as "extending the hand of kindness" through clean-ups to "bring home the message that Johannesburg cares about its people."[87] While employing the zero tolerance language of "hotspots" and "by-laws contravention," the press release depicts the operation as "restoring human dignity" and paternalistically educating inner city residents to "respect other people's rights."[88] Rather than "pushing people" such as street traders out of urban spaces, the city accepted that they are "making a living."[89] However, the inner city is also viewed as a place of "lawlessness and general disorder" from violent crimes to misdemeanours such as "loud music and cars being washed on the street," which are seen to necessitate regular street level operations to ensure compliance.[90] The focus on "general disorder" indicates how pacification is undergirded by the fear of insecurity, in which minor infractions are seen to belong to the same continuum as far more violent and destructive acts.

"Blitzing" also served an internal role within the COJ, being used to monitor and quantify the performance of officials. After identifying areas that were having problems with service allocation or by-law enforcement, a blitz would then be launched to serve notices ahead of police action or to fix infrastructure.[91] These blitzes earned points on a scorecard, which was used as the recording basis for performance bonuses—in turn providing a financial incentive to regularly launch operations.[92] This approach was controversial in the city with some officials arguing that these were unnecessary incentives for making people do their jobs, while others were in favour of retaining these bonuses. Beyond the pay imperatives, blitzes were psychologically gratifying. Taking part in blitzes was exciting for many officials who would in some cases have lunch packs ready for operations and in other cases bullet proof vests for raids on dangerous buildings.[93]

However, the reliance on blitzes from the early 2000s onwards was increasingly criticized as a reactive approach to disorder. For example, an article published on the city's official website argued against the:

> blitz mentality ... let things get so bad that you can no longer ignore them and then go in with all guns blazing. Thereafter fill the media with stories about how good we are, we cleared out X tons of waste from the area; we arrested Y number of known criminals, druglords and so forth; and we sent Z number of illegal immigrants back to where they came from.[94]

Officials interviewed for this article generally held the view that while blitzes were successful in both dealing with day-to-day security issues and conveying the sense of the state taking action, it was not a sustainable tactic for ordering the urban environment.[95] Blitzes are too "hit and run": traders "would run away" before returning after the authorities had left, along with cases of the JMPD tipping businesses off ahead of operations.[96] Furthermore, the blitzes had a marked spatial bias in which intensive force was aimed at the inner city, while tepid "Mickey Mouse" blitzes were launched elsewhere.[97] As a result, from 2009, blitzes began to fall out of favour within City government as a more systematic, "block-by-block" approach towards urban operations became the preferred terminology.[98] However, other institutions such as the JMPD still use the term as a general description for raids and expanded interventions. The term also appears to have been adopted within the private sector. A business proposal to establish a new improvement district in the inner city describes "the first phase …[which] will be to carry out a Blitz in the whole area [so that] place marketing initiatives can be introduced."[99]

The criticisms of blitzes were primarily tactical, based on their perceived inefficiency and crudity. But strikingly, even "sophisticated" alternatives to flooding the streets with state forces derive from the same historical matrix of warfare as blitzes. In particular, the City Safety Strategy, drafted in 2003, discussed a long-term program of consolidating order in a number of "areas important for … economic development" entailing programmes of "winning back the streets" through developmental initiatives and "focused surveillance" until the areas of order overlapped.[100] Underscoring the fact that this proposal was derived from previous counter-insurgency experience, this was defined as an "Oil Spot Geographically-Based Strategy."[101] The historical association of this term with Apartheid repression meant that its inclusion was criticized by other officials.[102] So it appears that while the methodology was maintained, the terminology was changed with a public information sheet about the Safety Strategy instead using the more neutral term "priority areas."[103] Johannesburg was not the only city to dust off old repressive blueprints as a guide to securing the urban future, as the oil spot concept was also used in the context of renewal plans in Cape Town.[104]

CONCLUSION

Although it is to be expected that these terms will persist in the institutional memory of city governance, it is nevertheless striking that a concept associated with the Apartheid state violence can be easily repurposed for urban development and regeneration in twenty-first century South Africa. Unlike in the past, where urban governance revolved around a white minority using law and violence to maintain strict spatial segregation, contemporary Johannesburg sees a democratically elected government implement a variety of schemes to manage and control the (at least officially) free circulation of people and spaces. As evidenced in blitzes launched against everything from organised crime to housing and overgrown parks, all areas of urban life are potential sites for often authoritarian state intervention—self-consciously planned with military precision and performed with a "human face." The state regards the urban spaces of the black and working class as needing regular shots of discipline, highlighting the continuation of racialised policing practice in the present (Steinberg 2011). This is evidenced by the tactic of blitzes, which overwhelms and surprises targets. As such, on the continuum of pacification it still relies more on coercion than winning consent.

Moreover, the blitz tactic also indicates the political economy of security. This economic core is what connects different practices and tactics of power across the history of Johannesburg, from air-raids in the 1920s to suppress insurrection, to a police tactic that derives its name from aerial bombing being used an everyday police tactic in early twenty-first century. Of course, this could just be written off as a case of just family resemblance, especially considering how radically South Africa has changed, not just in the last century, but in the last two decades. But as the theory of pacification points at the level of theory, these different activities are linked. The state is constantly linking war power and police power in the pursuit of economic security and social order. This is further indicated by how the blitz intersects with zero tolerance and counterinsurgency rhetoric—regardless of the tactical impacts of these concepts, it highlights the global dimensions of urban pacification, especially at the planning and conceptual levels- and in places where the class and racial aspects of domination are less overt than South Africa. The upshot is that in Johannesburg and elsewhere the logic of security entails that the state is engaged in endless wars against urban "lawlessness" and for order. While this is worked out in many different ways, within many different urban spaces, the concept of pacification allows us to see this as a continuum of rule and power. Perhaps most provocatively, it highlights the intrinsically authoritarian nature of capitalism, as state power is continually extended to create productive urban spaces and subject/consumers. Through various security and policing tactics and technologies, the state and capital attempt to fix complex urban environments into controlled and pacified cities—complete with docile labour forces and safe for consumerism and investment. Although the "blitz" term was particularly ubiquitous in Johannesburg, it is paralleled by many other types of urban police-military raid throughout the world. While they may fall under different names, they indicate that the domination of urban space through raids, planned like a military operation, advertised as humanitarian intervention and generally tested against the urban poor, is a dominant tactic of power in cities dealing with austerity, inequality, and simmering revolt.

NOTES

1 Stephen Graham, *Cities Under Siege: The New Military Urbanism* (London: Verso, 2010), xiv, 19, 114.

2. Oliver Belcher, "The Best-laid Plans: Postcolonialism, Military Social Science and the Making of US Counterinsurgency Doctrine," *Antipode* 44, no. 1 (2011): 258-263; Katharyne Mitchell "Zero Tolerance, Imperialism, Dispossession," *ACME: An International E-Journal for Critical Geographies* 10, no. 2 (2011): 293-312.
3. David Harvey, "The Political Economy of Public Space," in *The Politics of Public Space*, eds. Setha Low and Neil Smith (New York: Routledge, 2006); Harvey, "The Right to the City," *New Left Review* 53 (2008): 31.
4. Mark Neocleous and George Rigakos, "Anti-Security: A Declaration," in *Anti-Security*, eds. Mark Neocleous and George Rigakos (Ottawa: Red Quill Books: 2011), 15-22.
5. Mark Neocleous, " "A Brighter and Nicer New Life": Security as Pacification," *Social & Legal Studies* 20, no. 2 (2011): 191-208.
6. George Rigakos, " "To Extend the Scope of Productive Labour": Pacification as Police Project," in *Anti-Security*, eds. Mark Neocleous and George Rigakos (Ottawa, Red Quill Books: 2011): 57-84.
7. Sebastian Saborio, "The Pacification of the Favelas: Mega-Events, Global Competiveness and The Neutralisation of Marginality," *Socialist Studies/Études Socialistes* 9, no. 2 (2013): 130-145.
8. Mark Neocleous, " "A Brighter and Nicer New Life": Security as Pacification," 202-203.
9. Chris Ealham, *Anarchism and the City: Revolution And Counter Revolution in Barcelona 1898-1936* (Oakland: AK Press, 2010): 14.
10. Ananya Roy, Stuart Schrader, and Emily Crane, "Gray Areas: the War on Poverty at Home and Abroad," in *Territories of Poverty*, eds. A. Roy and E. Crane (Athens, GA: University of Georgia Press: 2015).
11. Roy, Schrader, and Crane, "Gray Areas."
12. Mike Davis, *City of Quartz: Excavating the Future in Los Angeles* (London, Verso: 1990): 260, 268.
13. Davis, *City of Quartz*, 274.
14. Saborio, *The Pacification of the Favelas: Mega-Events, Global Competiveness and the Neutralisation of Marginality*.
15. Christopher McMichael, "Sporting Mega-Events and South-to-South Security Exchanges: a Comparative Study of South Africa and Brazil," *The Hague Journal of Diplomacy* 8, no. 3-4 (2013): 313-332.
16. Raul Zibechi, *Territories in Resistance: A Cartography of Latin American Social Movements* (Oakland: AK Press: 2012): 193.
17. Raul Zibechi, *Territories in Resistance*, 192.
18. Escalating Identity, *Who is Oakland: Anti-oppression Activism, the Politics of Safety, and State Co-optation* (2012), 9-10. Accessed 30 July 2014. https://libcom.org/files/whoisoaklandsyn.pdf.
19. Michel Foucault, *Security, Territory, Population: Lectures at the Collège De France, 1977-1978* (New York, Picador: 2007); Graham, *Cities Under Siege*.
20. Graham, *Cities Under Siege*.
21. Dewald Van Rensburg , "Brave New Joburg: Operation Clean Sweep," *City Press*, November 24, 2013.
22. Greg Nicolson and Thapelo Lekgowa, "Operation Clean Sweep: Not Just a Clean-Up but a Purge of the Poor," *The Daily Maverick*, November 15, 2013, accessed July 29, 2014, http://www.dailymaverick.co.za/article/2013-11-15-operation-clean-sweep-not-just-a-clean-up-but-a-purge-of-the-poor/#.U9diTvmSwmM.
23. African Diaspora Forum, "The City of Johannesburg Continues to Deny More Than 6000 Hawkers The Right to Make a Living Johannesburg CBD," October 28, 2013, accessed July 14, 2014, http://adf.org.za/?q=node/.
24. Dewald Van Rensburg, "Charges Dropped Against Traders" Lawyer," *City Press*, December 6, 2013.
25. City of Johannesburg, *Mayoral Committee Minutes*, March 20, 2014.
26. Philip Harrison, interview by author, June 18, 2014.
27. City of Johannesburg, *Inner City Charter Regeneration Charter*, 2007, accessed July 29, 2014, http://www.joburg-archive.co.za/2007/pdfs/inner_city_regeneration_charter.pdf.
28. Karl-Heinz Frieser, *The Blitzkrieg Legend: The Campaign in the West 1940* (Annapolis: Naval Institute Press: 2005).
29. Steve Goodman, *Sonic Warfare: Sound, Affect, and the Ecology of Fear* (Cambridge, MA: The MIT Press: 2010): xiv.

30 "Traffic Blitz Puts Brakes on Law-Breakers, Australian Broadcasting Corporation," October 14, 2013, accessed July 29, 2014, http://www.abc.net.au/news/2013-10-14/traffic-blitz-puts-brakes-on-law-breakers/5020614; "7 Suspects from New York Arrested in Italy-US Mafia Drug Bust," February 11, 2014, accessed July 29, 2014, http://www.nbcnewyork.com/news/local/Italy-New-York-Mafia-Gambino-Bust-244854771.html.
31 Steve Niva, "Disappearing Violence: JSOC and the Pentagon's New Cartography of Networked Warfare," *Security Dialogue* June 44 (2013): 197; Belcher, "The Best-laid Plans," 260.
32 Mark Neocleous "The Dream of Pacification: Accumulation, Class War, and the Hunt," *Socialist Studies/Études Socialistes* 9, no. 2 (2013): 7-31.
33 Mark Shaw, *Crime and Policing in Post-Apartheid South Africa: Transforming Under Fire* (Cape Town, David Philips: 2002): 87; Hilary Janks, *Literacy and Power* (New York, Routledge: 2010): 161.
34 Hilary Janks, *Literacy and Power*, 161.
35 Jean Comaroff and John L. Comaroff, "Naturing the Nation: Aliens, Apocalypse and the Post-Colonial State," *Journal of Southern African Studies* 27, no. 1 (2001): 647; Michael Neocosmos, *From "Foreign natives" to "Native Foreigners": Explaining Xenophobia in Post-Apartheid South Africa: Citizenship and Nationalism, Identity and Politics* (Dakar, CODESRIA: 2006), 101.
36 Martin J. Murray, *City of Extremes: The Spatial Politics of Johannesburg After Apartheid* (Durham, Duke University Press: 2011), 153.
37 City of Johannesburg, "Mega Blitz for Mshenguville Street and Skate Park in Mofolo, Soweto," Press release, February 10, 2014; City of Johannesburg, "By-law Blitz Held in Region G," n/d.
38 D. Vigneswaran, *Territory, Migration and the Evolution of the International System* (Houndmills: Palgrave Macmillan, 2013): 116.
39 Wayne Minnaar, interview by author, September 4, 2013.
40 Charles Van Onselon, *New Babylon, New Nineveh: Everyday Life on the Witwatersrand 1886-1914* (Johannesburg: Jonathan Ball Publishers, 2001): 32.
41 Lindsay Bremner, *Writing the City into Being: Essays on Johannesburg, 1998-2008* (Johannesburg, Fourthwall Books: 2010), 175; *Union Government Police Report of the Commissioner 1919* (Pretoria: Government Printers, 1921), 18.
42 Tyler Wall, "Unmanning the Police Manhunt," *Socialist Studies/Études Socialistes* 9, no. 2 (2013): 43.
43 Baruch Hirson, "The General Strike of 1922," *Searchlight South Africa* 3, no. 3 (1993): 61-94.
44 Irene Mafune, interview with author, July 16, 2013.
45 P. Frankel, "The Politics of Passes: Control and Change in South Africa," *The Journal of Modern African Studies* 17, no. 2 (1979): 207.
46 L. Schuster, "Pretoria's Bid for Hearts and Minds," *The Christian Science Monitor*, May 11, 1988.
47 Phillip Bonner and Noor Nieftagodien, *Alexandra: A History* (Johannesburg: Wits University Press, 2008), 386.
48 Andy Clarno, "Rescaling White Space in Post-apartheid Johannesburg," *Antipode* 45, no. 5 (2013): 1204.
49 Li Pernegger, "The Agonistic State: Metropolitan Government Responses to City Strife Post-1994," in *Urban Governance in Post-Apartheid Cities: Modes of Engagement in South Africa's Metropoles*, eds. Christoph Haferburg and Marie Huchzermeyer (Stuttgart: Borntrager Science Publishers, 2015), 61-79.
50 Margot Rubin, "Courting Change: the Role of Apex Courts and Court Cases in Urban Governance: a Delhi-Johannesburg Comparison" (PhD diss., Wits University PhD thesis, 2013), 86.
51 Philip Harrison, interview by author, June 18, 2014.
52 Richard Norton, *Feral Cities, Naval War College Review*, Autumn (2004): 97-106.
53 Lindsay Bremner, *Writing the City into Being: Essays on Johannesburg, 1998-2008* (Johannesburg: Fourthwall Books, 2010): 154.
54 Barbara Lipietz, " "Muddling-through": Urban Regeneration in Johannesburg's Inner City," (paper presented at N-Aerus Annual Conference, Barcelona, 2004).
55 City of Johannesburg *2030: Strategy* (2002), 121.
56 City of Johannesburg *2030 Strategy* (2002), 124-123.
57 City of Johannesburg *2030 Strategy* (2002), 124-123.

58   Loic Wacquant, *Punishing the Poor: The Neoliberal Government of Social Insecurity* (Durham, Duke University Press: 2008).
59   Philip Harrison, interview by author, June 18, 2014.
60   Rubin, *Courting Change*, 33.
61   Harvey, *The Right to the City*.
62   Lipietz, Muddling Through, 4.
63   Philip Harrison, interview by author, June 18, 2014; Rubin, "Courting Change," 91.
64   Dominique Malaquais, "Anti-Teleology," in *African Cities Reader: Mobilities and Fixtures*, eds. N. Edjabe and E. Pieterse (Vlaeberg: Chimurenga and the African Centre for Cities, 2011): 23.
65   Carolyn Kihato " "Here I am Nobody": Rethinking Urban Governance, Sovereignty and Power," in *African Cities Reader: Mobilities and Fixtures*, ed. N. Edjabe and E. Pieterse (Vlaeberg: Chimurenga and the African Centre for Cities): 70-77.
66   Kihato, "Here I am Nobody"; Philip Harrison, interview by author, June 18, 2014.
67   Philip Harrison, June 18, 2014.
68   City of Johannesburg, *Joburg 2040 Growth and Development Strategy* (2011), 76.
69   Ibid., 77.
70   Jonny Steinberg, "Crime Prevention Goes Abroad: Policy Transfer and Policing in Post-Apartheid South Africa," *Theoretical Criminology* 15, no. 4 (2011): 349-364.
71   Philip Harrison, interview by author, June 18, 2014.
72   Nazria Cachalia, interview by author, September 20, 2013.
73   Associated Press, "S. Africa Crime Sweep Nets 4,593," June 28, 1991, accessed July 31, 2014, http://articles.latimes.com/1991-06-28/news/mn-1438_1_crime-sweep-police.
74   Bremner, *Writing the City into Being*, 222.
75   Sapa, "Sixty Caught in Johannesburg Crime Blitz," *iolnews*, December 19, 1999, accessed July 31, 2014, http://www.iol.co.za/news/south-africa/sixty-caught-in-johannesburg-crime-blitz-1.23713#.U9n0_vmSwmM.
76   Sapa, "1,900 Arrested in National Crime Blitz," *iolnews*, March 31, 2000, accessed July 31, 2014, http://www.iol.co.za/news/south-africa/1-900-arrested-in-national-crime-blitz-1.33198#.U9n3NPmSwmM.
77   Tony Samara, "State Security in Transition: The War on Crime in Post Apartheid South Africa," *Social Identities* 9, no. 2 (2003): 288.
78   C. Bhengu, "South Africa: Apology to Those Wrongly Detained in Crime Blitz," *AllAfrica.com*, March 22, 2000, accessed July 31, 2014, http://allafrica.com/stories/200003220145.html.
79   Sapa, "1,900 Arrested in National Crime Blitz."
80   Neocosmos, *From "Foreign natives" to "Native Foreigners,"* 101.
81   G. Gifford, "Hotel From Hell Shut Down in Hillbrow Blitz," *iolnews*, November 22, 2001, accessed July 31, 2014, http://www.iol.co.za/news/south-africa/hotel-from-hell-shut-down-in-hillbrow-blitz-1.77373.
82   Davis, *City of Quartz*, 267.
83   Sapa, "Jhb gets a Clean-Up Blitz," *News24.com*, March 20, 2001, accessed July 31, 2014, http://www.news24.com/xArchive/Archive/Jhb-gets-a-cleanup-blitz-20010320.
84   COHRE, *Any Room for the Poor: Forced Evictions in Johannesburg* (Johannesburg: Centre for Housing Rights and Evictions, 2005): 42.
85   Ibid., 42, 45.
86   City of Johannesburg, *Inner City Committee—Integrated Inner City Safety and Security Action* Plan, 4.1, 4.28.
87   City of Johannesburg, "Extending the Hand of Kindness," March 4, 2008, accessed July 31, 2014, http://jborg.ubik.net/index.php?option=com_content&task=view&id=2252&Itemid=254.
88   City of Johannesburg, "Extending the Hand of Kindness."
89   Irene Mafune, interview by author, July 16, 2013.
90   Ibid.
91   Greg Daniels, interview by author, August 19, 2013.
92   Philip Harrison, interview by author, June 18, 2014.
93   Ibid.
94   Neil Fraser, "Some Easter Musings on the City," April 10, 2007, accessed July 31, 2014, http://www.joburg.org.za/index.php?option=com_content&id=484&Itemid=.
95   Nazira Cachalia, interview by author, September 20, 2013.
96   Greg Daniels, interview by author, August 19, 2014.

97 Phil Harrison, interview by author, June 18, 2014.
98 Greg Daniels, interview by author, August 19, 2014.
99 Maboneng Improvement District, Draft Business Plan (2014), 26.
100 City of Johannesburg, *City Safety Strategy: Executive Summary* (2003), 10.
101 Ibid., 10.
102 Nazira Cachalia, interview by author, June 6, 2013.
103 City of Johannesburg, *Inner City Charter Regeneration Charter* (2007): 2.
104 Samara, "State Security in Transition," 306.

Part 3
PACIFICATION AND POLICING THE SOCIAL ORDER

Pacification theory
An empirical test[1]

Aysegul Ergul & George S. Rigakos

It is increasingly becoming accepted wisdom that the widening gap between rich and poor is a significant public policy concern.[2] The 'gap' issue has been embraced by liberal thinkers as one of equity,[3] by venture capitalists as one of sufficient consumption and the spectre of economic stagnation[4] and by security experts as one of national stability. Witnessing falling real wages alongside soaring profits, of course, comes as no surprise to critical political economists [5] who have long held that capitalism produces gross disparities in relative wealth and income. In Marx and Engel's terms "[s]ociety as a whole is more and more splitting up into two great hostile camps, into two great classes directly facing each other — Bourgeoisie and Proletariat." [6] In the United States, resultant political discussions about the decline of the 'middle class' have turned to a more progressive income tax system as a solution[7] that might act as a leveler for inequality and a method by which worker insecurity may be alleviated. Yet, worker insecurity is of significant benefit to capitalists both domestically and internationally.[8] After all, worker uncertainty and exploitation is considered 'productive' for the economy as it significantly tempers demands and drives down real wages. In this way, we might say that an *insecure* workforce is one important step toward a *pacified* workforce. It is therefore the nature of capitalism to engage in warfare (both open and subterranean) against its workers in order to produce a consistent, beneficial insecurity: a state of anxiety that can only ostensibly be satiated by consumption. In fact, during the economic bubble of the late 1990s and just ahead of the Great Recession, U.S. Federal Reserve Chair Alan Greenspan argued that "growing worker insecurity" played a pivotal role in workers' having stopped asking for wage increases which was by extension beneficial for US capital.[9] Of course, this insecurity took place during a period that witnessed a massive rise in corporate profits, a decline in real income[10] and consistent increases in public and private policing.

In this paper, we investigate the relationships between four resilient trends: (1) the consistent erosion of union-membership; (2) an increase in income polarization and inequality; (3) a dramatic resurgence in popular protest; and (4) a steady rise in public and private policing employment. We analyze the relationship between these variables in the context of a theory of pacification which argues that "policing", broadly defined, has both historically and contemporaneously been designed to "make workers productive"[11] by "fabricating a social order"[12] that seeks to protect private property relations and promote capital accumulation. In this sense, we treat total policing employment as an empirical barometer of bourgeois insecurity conditioned by two elements of Marxian political economy: (1) relative deprivation (income inequality), and (2) the rise of an industrial reserve army, or manufacturing unemployment through successive deindustrialization. We also examine how both worker exploitation (as measured by surplus value) and labour militancy (as measured by strikes and lockouts per 100,000 population) interact with declining union membership. Our goal, in simplest terms, is to empirically test the central tenets of pacification theory.

This paper is organized into five sections. The first section introduces the reader to the basic theoretical tenets of pacification and policing, and identifies a number of assertions that are operationalized in a manner that allows us to statistically test their veracity against our data. The general relationships measured in this paper all revolve around policing employment as a contemporary barometer of bourgeois insecurity, yet it is important to note that this is only one aspect of the broad project of police and its relationship to capital. The next section on methodology lays out the variables and second-order variables, calculated by us, such as surplus value and cumulative deindustrialization. The following three sections explain and contextualize the results of our analyses.

POLICING AS PACIFICATION

Pacification is the continuum of police violence upon which the fabrication of capitalist order is planned, enforced and resisted. The *Oxford English Dictionary* defines pacification as a state or sovereign action that attempts "'to put an end to strife or discontent' and 'to reduce to peaceful submission' a rebellious population." Neocleous argues that the Dictionary's reference to the *Edicts of Pacification* of 1563, 1570 and the *Edict of Nantes* in 1598 as the first instances of the usage of the word 'pacification' are important because "they are the point of departure for the period in which the insecurity of bourgeois order had to be secured."[13] In this context, a politics of security, a need to fabricate an order necessary for the functioning of the bourgeois state becomes central to the development of liberal philosophy both domestically and imperially,[14] and it is at this point that a science of police (*Polizeiwissenschaft*) becomes the most suitable means to proactively produce a new order.

This police science has a rich analytic history, going back to the 17th century, tied to the development of a governmental need to order populations and to fabricate conditions conducive to capitalist accumulation. Much of what we now come to understand as policing is based on Enlightenment thought concerning the best organization of populations for the "welfare of all" and the maximization of wealth either for an absolutist monarchy or, later, as part of the market dynamics of a liberal state. These organizing principles of police and capital have historically revolved around an intentional class-based ordering, including the proper establishment of work-houses, the best method of keeping accounts of persons and goods as well as their movements, the formation of a pauper police, and the systematic categorization of worthy and unworthy poor. These innovations in police thinking played a prominent and pronounced role in the establishment of the 19th century constabulary that the English-speaking world, including the United States, has inherited. The "new police" of London were a bourgeois innovation that were made necessary by an unregulated migration of "masterless men" and other "vagabonds" to the emerging industrial heartlands of England. The police were set up to methodically inculcate a wage-labour system that made capital accumulation possible and more predictable.

The great police thinker Patrick Colquhoun sought to set up a system of enforcement as a method to regulate the compensation of labour conducive to capital accumulation.[15] At the heart of 19th century imperial England, Colquhoun set about creating an experimental police that would replace "chips and perquisites" of all types among dock workers and that would enforce a dress code and system of inspection that would "eliminate pilfering" altogether. Colquhoun's methods were centred on the enforcement of wage labour and his success was catalogued empirically. As Rigakos[16] notes, Colquhoun "clearly realized that social control… was geared to the benefit of a particular class of property holders" and that this was "consistent with his emphasis on managing the various classes of persons who he said threatened commercial interests." Thus, the main target and concern of police has from its inception been the working class and the poor. "That is, its mobilizing work was the mobilization of work."[17] Colquhoun, however, was also an imperialist. A former Virginia colonist and Loyalist, he even raised an army out of Scotland to assist in putting down the American rebellion. Like other police intellectuals, Colquhoun focused on both imperial planning by means of fabricating a wage-labour system and putting an end to domestic instability.[18] He was, in the strictest sense, both a theorist and practitioner of pacification.

In the American context, labour unrest and its policing also has a very long history. Like the British context, there were experiments with private policing such as the Coal and Iron Police, who worked directly for industrial interests and were often brutal in their methods of strike-breaking and unscrupulously infiltrating and undermining worker associations.[19] When private security companies such as the notorious Pinkerton Detective Agency[20] proved too controversial and local guardsmen proved too unpredictable by galvanizing further resistance,[21] states across the union began to move to state-level law enforcement in an effort to create a more centralized, less locally dependent, and 'professional' service.[22] (In the same way it was clear to workers and political agitators in the 19th century

London, it was not lost on American labour activists of the day that the legislative move towards the use of state "troopers" and "rangers" was a direct threat to their ability to mobilize. At every step, pacification anticipates resistance. It is within this historical backdrop, both domestic and international, that the interconnected role of police and capital can be viewed as part of a large-scale project of pacification. Thus, there are at least five tenets that may be distilled from current thinking about pacification that are useful to our study:

1) **Public-private.** Given their institutional interchangeability yet identical targets of enforcement, to rigidly distinguish between public versus private forms of policing is to further reify a false binary that obfuscates far more than it reveals.[23] Put another way: "The public sphere does the work of the private sphere, civil society the work of the state. The question is therefore not 'public versus private' or 'civil society versus the state', but the unity of bourgeois violence and the means by which pacification is legitimized in the name of security."[24] In the context of both pacification and a Marxian political economy, it makes no sense to operationalize public and private policing separately and so, in this paper, we *operationalize public and private policing employment into one variable.*

2) **Inequality.** The more capitalism *naturally* matures the more unequal the distribution of wealth. Adam Smith admitted as much but defended the emergent class distinctions that sprang from early capitalism by unapologetically arguing that the "accommodations" of "an industrious and frugal peasant" always "exceed[ed] that of many an African king, the absolute master of the lives and liberties of ten thousand naked savages."[25] He held that *absolute* poverty is reduced wherever capitalism flourished. Marx, however, pointed out that poverty was *relative* rather than absolute arguing that since our gains and possessions "are of a social nature, they are of a relative nature."[26] Despite the fact that living conditions may have improved for the lowest rungs of society, they improved much more significantly for the bourgeoisie whose source of wealth was directly tied to the exploitation of workers. If policing is aimed at the fabrication of an order that seeks to promote capital accumulation and the valorization of private property, then the larger the threat to that order the greater the aggregate need to secure it. Rising inequality is a threat to the capitalist order because it amplifies relative deprivation. For us, this means that, over time, unfettered capitalist systems become more and more unequal. This inequality, if not addressed by other means, must necessarily occur alongside more and more policing. *Inequality will positively correlate to total policing.*

3) **Surplus value.** Long before the Thames River Police and in the preceding pre-capitalist economic epochs some form of coercion has always been required to realize a surplus.[27] The historical and institutional goal of police science, both in terms of planning and enforcement, is to produce an environment conducive to the promotion of capital accumulation, to make workers productive. This is accomplished by facilitating the practice of exploitation necessary for the functioning of the capitalist system. Exploitation, in Marxist terms, is unpaid labour time or surplus value: the amount of time that a worker works without getting paid and for which the capitalist realizes a surplus.[28] A system of police is vital to the extraction of surplus value because it is based on the use and threat of coercive force. *Surplus value will positively correlate to total policing.*

4) **(De-)Unionization.** There is ambivalence among Marxist thinkers about the relative role of unions with respect to the revolutionary goals of the proletariat. Gramsci argued that unions "cannot be the instrument[s] for a radical renovation of society."[29] Luxemburg lamented that unions suffered from a "bureaucratism and a certain narrowness of outlook" because their goal was to ameliorate and resolve class conflict as much as possible.[30] Trotsky chastized "the reformist bureaucracy and the labour aristocracy who pick the crumbs from its banquet table"[31] and derisively dubbed this emerging labour aristocracy the "lieutenants of capital."[32] By 1872, complained in his speech to the General Council of the International Workers' Association that "[t]rade unions are praised too much; they must in the future be treated as affiliated societies and used as centers of attack in the struggle of labour against capital."[33] In sum, the general position toward trade unions by Marxists is that unions alone cannot be the vehicles for the radical transformation of the social relations of production because, by their very nature, trade unions do not seek to unleash the war between the bourgeoisie and the proletariat but rather act to keep the peace: to behave, as Trotsky put it, as policing agents for capital. If we are to believe that unions are indeed agents of capital, then they also acquire a police function in society and so we should see that more union membership per capita will result in lower police employment numbers, and the inverse should also be true. *Unionization will inversely correlate to total policing.*

5) **The industrial reserve army.** It is fair to say that there has been no greater preoccupation among the police scientists of the Enlightenment than with that of the idle, the indigent and the poor. To a large extent, planning for control over and policing of vagabonds and masterless men, the creation of policy to discern between the deserving and undeserving poor, and motivating people classified in these categories to become productive have

guided much thinking about the proper use and deployment of police.[34] Marx also had much to say about this reserve army of workers. For they act as both a pressure release on the demands of workers by lowering expectations – lest they be replaced by the unemployed – and also appear as a threat to the system itself by acting as a ready reservoir of agitated revolutionaries. Marxian historians have offered significant insights into how the emergent bourgeois state made it a central organizing mission to pacify this body of transient workers through forced migration, immigration policy and the use of public and private police to crush revolutionary agitation.[35] Both in the formative logics of bourgeois systems of police and within the radical philosophy of revolutionary politics, the industrial reserve army plays a pivotal role. Much of the revolutionary fervor of the 19th century around the unemployed persists today. Strikes and lockouts may act as catalysts for revolutionary actions[36] and union agitation becomes paramount; yet, with ever-decreasing union membership it is likely that strikes and lockouts will also become less frequent. Thus, we hypothesize: *Unionization will positively correlate to strikes and lockouts* while *manufacturing unemployment and/or cumulative deindustrialization will positively correlate to total policing.*

METHODOLOGY

We have now laid out five general tenets of Marxian political economy, generally, and pacification more precisely. We have operationalized these tenets in a hypothetico-deductive manner for the purpose of statistical testing. Our variables are as follows: (1) total policing employment per 100,000 population which is the sum of (a) public police employment per 100,000 population and (b) private security employment per 100,000 population; (2) inequality; (3) surplus value; (4) union membership; (5) manufacturing unemployment, or cumulative deindustrialization (which is the cumulative annual difference of manufacturing job losses subtracted from manufacturing job gains); and (6) strikes and lockouts.

In this article, we test the relationship between these variables for 45 primarily northern industrialized countries for the snapshot year 2004[37], and U.S. time-series data from 1972-2009. In doing so, we aim to test our hypotheses both internationally and domestically. Our statistical analysis consists of correlations and comparisons of means. We restrict ourselves to this level of statistical analysis for two reasons. First, there is no epistemic basis for classifying a dependent variable among these measures according to the tenets of Marxian political economy. In fact, our theoretical analysis has demonstrated that these relationships are conditional and relational. Second, this is an exploratory study and so it is more than sufficient at this stage to work toward building a model based upon relational connections.

Public police and private security data are added to make up our measure of **total policing**. Public police and private security statistics utilized in our international comparative study are derived primarily from publicly available EU sources[38] while U.S. time-series data is derived from census, FBI and labour statistics.[39] To calculate overall **unionization** internationally, we primarily used data from the World Values Survey (WVS).[40] The unionization variable utilized in this analysis is based on the following question from the WVS: "Do you belong to labour union?" We also calculated union-membership as a percentage of the population by using data from the OECD Stat Extracts,[41] European Industrial Relations Observatory Online[42] and population data from the World Development Indicators. This data is used as a substitute for missing reports from the WVS. The U.S. time-series data is derived from three major sources: (1) "Union Membership Trends"[43]; and (2) News Releases prepared by the BLS; and (3) "Union Membership, Coverage, Density, and Employment by State and Sector, 1983-2011."[44]

The most widely used measure of **income inequality** in existence today is the Gini coefficient.[45] With few exceptions we employed the World Income Inequality Database (WIID) compiled by the World Institute for the Development of Economic Research (WIDER) online. We use a three-year average Gini coefficient leading up to 2004. For the U.S. time-series data, we used the Income Inequality database, Earnings and Poverty Reports.

Perhaps no other Marxist economic concept has received as much analytic attention as **surplus-value**. This is perfectly understandable given that it is the fundamental measure of exploitation, the source of wealth for the bourgeoisie, and the engine of the capitalist mode of production. Measuring surplus-value, however, is no easy task. Bourgeois economic statistics do not directly capture what Marx meant by the rate of surplus-value in his calculation:

$$SV = \frac{\text{volume of surplus produced}}{\text{variable capital expended (labour)}}$$

While there have been diligent attempts to proximate the rate of surplus value using existing economic measures,[46] the simplest and most widely applicable measure adopted by Marxian scholars,[47] especially for facilitating international comparisons is:

$$SV = \frac{(\text{gross value added} - \text{total manufacturing workers' earnings})}{\text{total manufacturing workers' earnings}}$$

We adopted this calculation in our analyses.[48] The U.S. time-series data for surplus-value was retrieved from publicly available census and labour statistics.[49]

The data for **strikes and lockouts**, and **manufacturing unemployment** were mostly obtained from the International Labor Organization (ILO) online. We re-calculated strikes and lockouts as a rate per 100,000 population in order to make international comparisons. In our U.S. time-series, we replaced our manufacturing unemployment variable with a new "cumulative deindustrialization" measure. The reason for such an alteration derives from the statistical restrictions that the use of manufacturing unemployment data generates in a longitudinal analysis. The Current Population Survey produces "unemployment by industry" data by asking for the identification of the last job that the persons participating the survey held. This poses a problem for our longitudinal analysis because people who were laid-off from manufacturing sector employment could be hired and again laid-off from jobs in another sector prior to the survey. Considering that displacement is a major issue for manufacturing sector employees,[50] manufacturing unemployment as a variable loses its reliability over time. Therefore, in our analysis, we have created a measure of *cumulative deindustrialization*: which, as mentioned, is the net change in deindustrialization calculated by subtracting the annual "job destruction" or job losses from "job creation" or job gains. It is a cumulative measure because we add the number of each year's net change to the following year. The data for "job destruction" and "job creation" as well as job losses and job gains which we add as a measure of **cumulative deindustrialization** is retrieved from Business Dynamics Statistics released by the U.S. Census Bureau, and Business Employment Dynamics Statistics produced by the BLS. National level data concerning **strikes and lockouts** is obtained from Economic News Releases produced by BLS.

INEQUALITY

Today, advocates of global "free market" capitalism believe neo-liberalism is the stimulus of growth and development. They advocate openness to trade and investment in order to spur economic growth that in turn raises income levels and the standard of living. Developing countries that open their markets and liberalize their economies therefore grow faster which in turn leads to the narrowing of income differences.[51] In post-Soviet systems, for example, neo-liberal reformers have concluded that in countries where privatization, trade liberalization, and labour deregulation policies were adopted quickly and aggressively, GDP grew the fastest and people saw the greatest changes in their standard of living.[52]

Scholars on the other side of the spectrum, however, have demonstrated that income inequality has continued to rise rather than decline under neoliberal reforms.[53] They find that world income inequality has actually increased over the past two to three decades both between and within countries.[54] Pay inequality within countries was stable or declining from the early 1960s to the early 1980s, but it has increased sharply and continuously across the globe since.[55] It is at this point that the ghosts of Smith and Marx most decidedly haunt the present. The proportion of the world's population living in extreme poverty, critical scholars must concede, has fallen precipitously in the last three decades especially with the economic rise of China and India. In this way, Smith is clearly vindicated: "unfettered" capitalism does indeed elevate the lowest out of absolute poverty and they are much wealthier today than they were before. But in 2010 as in 1850, the issue for Marxists remains relative deprivation. On this measure – comparing the richest to the poorest – income polarization has increased markedly. The income gap between people living in the top fifth of the richest countries and those living in the bottom fifth was 30:1 in 1960, 60:1 in 1990, and 74:1 in 1997. In 2005, the Human Development Report stated, "[t]he world's richest 500 individuals have a combined income greater than that of the poorest 416 million." [56] In the same year the 2.5 billion people, or 40 per cent of the world's population, that lived on less than two dollars a day accounted for five per cent of global income while the richest ten per cent, almost all of whom lived in high income countries, accounted for 54 per cent of global income.[57] According the World Institute for Economic Research, the richest two per cent of adults in the world owned more than half of global household wealth while the poorer 50 per cent of the world's adults owned

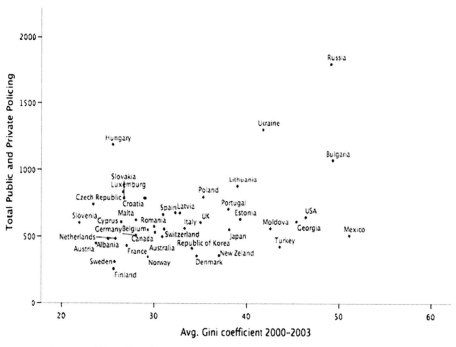

Graph 1. International Snap-shot: Total Policing by Inequality

Pearson r= .344, n=43, p<.05

barely 1% of global wealth.[58] Since the late 1960s U.S. income disparity has become the widest among all major industrialized nations.[59] The Gini co-efficient rose from 38.6 in 1968 to 46.8 in 2009 to 46.9 in 2010.[60]

When we tested our hypothesis, that – *Inequality will positively correlate to total policing*, we found that in our international sample inequality as measured by the Gini coefficient was statistically significantly correlated to total policing employment (r^2=.344, n=43, p<.05, see Graph 1). Our time-series analysis of the U.S. (from 1972 to 2009) demonstrated that there is an even stronger positive correlation between total policing and inequality (r^2= .940, n=36, p<.001, see Graph 2). Indeed, the strength of the association in the US is particularly striking.

Marx's notion of relative deprivation describes an increasing sense of discomfort and dissatisfaction among workers with the rise of domestic inequality. We have interpreted this as part of Marx's general theory of exploitation and, of course, proletarian immiseration. As a barometer of bourgeois insecurity, therefore, a positive relationship between policing and income inequality confirms the Marxist position about the class-based role of policing[61] above and beyond its direct relationship to labour. Our analysis demonstrates that increasing inequality in the U.S. has risen in almost synchronous lockstep with a rising body of public and private policing agents in the last four decades.

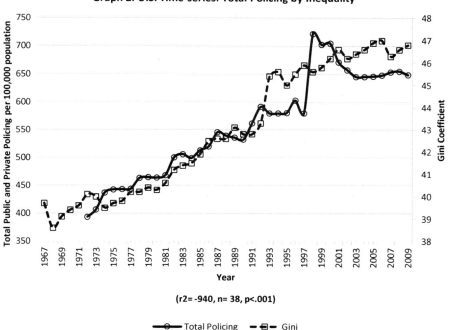

Graph 2. U.S. Time-series: Total Policing by Inequality

($r2$= -940, n= 38, p<.001)

SURPLUS VALUE

Surplus value is both the aggregate effect and the driving engine of capitalism. It is the unpaid labour-time a capitalist must steal from a worker in order to create profit or margin. In Marxian terms, capitalism does not exist without surplus value. As a basic maxim of capitalism one must drive down labour costs to increase surplus, to maximize profits. As inequality rises alongside declining unionization (as we shall see in the next section) it makes perfect sense to presume that this will facilitate the further economic exploitation of manufacturing workers. Job losses breed insecurity which fuels wage concessions. The more pacified the labour force, the more 'productive' it is in the eyes of capital. But this widespread unease, immiseration and exploitation produce unease on the part of the bourgeoisie which necessitates more policing.

When we tested our hypothesis – *Surplus value will positively correlate to total policing*, our analysis of the U.S. showed a very strong positive correlation (r^2= .855, n=39, p<.001, see Graph 3) between the rate of exploitation (the rate of surplus value) and policing. We did not observe a similarly significant relationship in our international study though the direction was also positive. This difference between the U.S. and the rest of the world may be explained by intervening

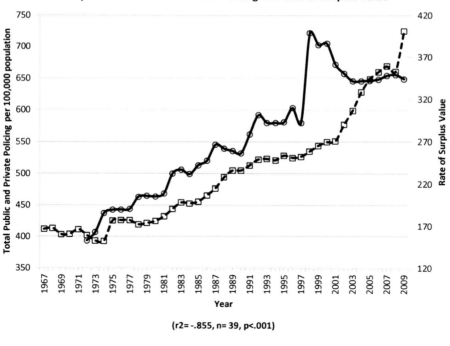

Graph 3. U.S. Time-series: Total Policing and Rate of Surplus Value

($r2$= -.855, n= 39, p<.001)

variables such as the relatively more well-developed social safety net in OECD countries, especially in Europe[62] that may ameliorate the need for more policing in cases of higher exploitation through social welfare initiatives.

INDUSTRIAL RESERVE ARMY, DE-UNIONIZATION AND DE-INDUSTRIALIZATION

The industrial reserve army (IRA) plays an important role in Marxian political economy – it acts as a calibrating body that drives down wages in times of economic downturn through heightened competition for work. The IRA (or sometimes surplus population[63]) helps moderate the costs of labour through a heightened competition for jobs. Capitalists can thus get away with further intensifying the exploitation of workers and depressing their wages. But the reserve army also threatens the entire capitalist structure by acting as a reservoir for the desperate and dispossessed proletariat who stand ready to seize the means of production. While Marx[64] was quick to differentiate himself from the classic political economy of Adam Smith who "fetishized" the production of "vendible commodities" as supremely productive, he nonetheless follows a similar course by all but dismissing the service sector economy from his analysis and repeatedly linking "surplus-value" – the engine of wealth creation under capitalism – to manufacturing.[65] For Marxian political economy, therefore, the IRA plays a key revolutionary role and is, by extension, an indirect threat that the bourgeois state must secure itself against. Understanding and measuring the IRA, that mass of workers made redundant by the rising organic composition of capital, and the *rate of exploitation* they indirectly help foster are thus central Marxian concerns. If we are to attempt to operationalize Marx's notion of the industrial reserve army we need to take stock of how he describes its function. In Chapter 25 of *Capital*, Marx likens the IRA to a law-like, supply-and-demand, regulator of labour costs for capitalists. The more efficient that production becomes through the use of machinery (i.e. the tendency of the organic composition of capital to rise) the more workers become redundant, the more plentiful their supply, the lower the wages they are willing to work for, the more likely they are to be exploited in less efficient industries until they are once again thrown out of work. The most important mechanism used by workers to offset this exploitative relationship is collective bargaining.

Although the strength of trade-unions reached a peak point in the post-War period both in their social and political impact and in the overall membership, it was not long before this process was systematically reversed. Industrial countries have been witnessing a steady decline in trade-union membership rates over the last three decades.[66] In the existing literature, this global decline in union membership is explained through (1) cyclical, structural, and institutional factors[67]; (2) individual membership decisions (personal and workplace characteristics, social environment)[68]; (3) product market competition[69]; and (4) changes in normative orientation from collectivism towards individualism.[70] These changes, however, must be viewed within the overarching context of the intensification of production in the wake of a paradigm shift away from Keynesianism to neo-liberalism at the beginning of 1980s. Deregulation, decentralization and extensive privatization are the main characteristics of neo-liberalism and they have generated a shift toward the decentralization of bargaining, labour market deregulation and the flexibilization of production in capital-labour relations.[71] As a result, a hospitable Keynesian postwar environment has been replaced with the enactment of discouraging, if not hostile, labour legislation and new regulations concerning industrial relations since the 1980s. [72]

Newly elected conservative governments passed legislation to "tame trade-unions" which they perceived as the cause of low productivity. Changes made to the legal frameworks governing unions included the banning of closed shops, promoting individual bargaining over collective bargaining, decreasing legal immunity (available to unions for damages in their activities), discouraging recruitment and strikes, and translating trade-union services into public goods.[73] The introduction of independent unemployment insurance funds in Ghent countries, like Finland and Sweden, or the replacement of voluntary but publicly supported unemployment insurance managed by unions with statutory regulations in Norway and the Netherlands has also generated a decline in union density because the connection between earnings-related unemployment benefits and union-membership has gradually broken down.[74]

When we tested our hypothesis – *Unionization will inversely correlate to total policing*, in our international study, union membership seemed to be unrelated to policing employment for all countries. But when post-USSR states were removed from the sample a statistically significant inverse relationship between private security ($r^2= -.324$, n=38, p<.05) as well as total policing employment ($r^2= -.426$, n=38, p<.01) and unionization appeared (see Graph 4). Indeed, post-USSR states exhibit a contrary positive relationship between total police employment and unionization.[1] This finding provides empirical evidence for the claim that unions may actually provide a surrogate policing function for capital in western nations. That is, a stronger union presence lessens the necessity for more policing. This is particularly evident among northern European (and Ghent countries) where the average unionization rate is 25.6%, the highest by far among all regions, but the average total policing rate is 453.4, the lowest among all regions. In former USSR countries, on the other hand, a high unionization rate (12.5%) coincides with more policing, particularly public policing (620.6) as the massive post-totalitarian apparatus has been largely maintained in the form of new protection rackets.[75] Eastern Bloc states have also had to deal with similarly bloated post-totalitarian security structures but the average policing rate is 780.5 while the unionization rate is 10.9%.

---

1   Although not a statistically significant one ($r^2=.632$, n=7, p=*ns*)

The steady decline in global unionization rates has been coupled with a precipitous decline in manufacturing employment globally.[76] The U.S. has also been experiencing a steady decline in the union membership[77] and the manufacturing sector's share of overall employment since its all-time peak in 1979.[78] The job losses in the manufacturing sector have been even more substantial since 2000. Following the 2001 recession, employment in manufacturing fell by 17%, and by the end of 2007 "had edged down further"[79] and has continued to fall subsequently.

When we tested our hypothesis – *Cumulative deindustrialization will positively correlate to total policing*, we found that the upward trend in industrial job losses in the US has coincided with rising total policing employment numbers (r^2= .705, n=32, p<.001). Graph 5 illustrates this trend line since 1972. Again, as in our study of the relationship between the industrial reserve army and total policing internationally (r^2=.462, n=34, p<.01) more industrial unemployment coincides with more policing.

Lastly, when we tested our hypothesis – *Unionization will positively correlate to strikes and lockouts while both unionization and strikes and lockouts will inversely correlate to total policing*, in the U.S. time-series, we found that both unionization (r^2= .932, n=39, p<.001) and strikes and lockouts (r^2=.818, n=39,

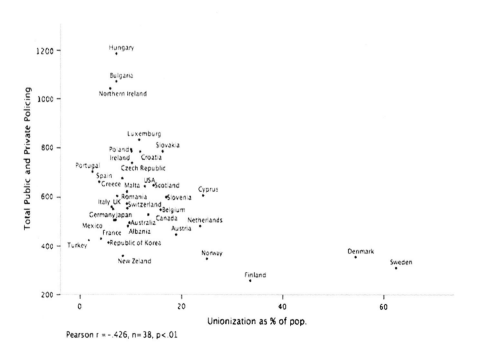

Graph 4. International Snap-shot: Total Policing by Unionization as % of Population

Pearson r = -.426, n= 38, p<.01

p<.001) are statistically significantly inversely correlated to total policing (see Graph 6). That is, as unionization has declined so have strikes and lockouts while the number of total police per 100,000 population have increased. In fact, while total policing employment has climbed by 64.8% from 1972 to 2009, strikes and lockouts have declined by 99.2% and the overall unionization rate has dropped by 55.4%. A pacification approach to unions would hold that these organizations play a surrogate policing role. Certainly, our international study reinforces these statistically significant findings (r^2= -.426, n=38, p<.001) – as unionization decreases, policing increases. Similarly, as unionization decreases so does the frequency of strikes and lock-outs decrease both internationally and in the USA (r^2=.923, n=50, p<.001). Not surprisingly, as both unionization and strikes and lockouts have decreased consistently since the early seventies so has deindustrialization increased. That is, as cumulative job losses in the industrial sector mount, so does unionization wither (r^2= -.823, n=32, p<.001) and strikes and lockouts decrease (r^2= -.691, n=32, p<.001, not shown).

CONCLUSIONS

At the outset of this paper we produced five testable hypotheses that were gleaned from our reading of Marxian political economy and studies of pacification. Our aim was to examine the explanatory power of these approaches for assessing

Graph 5. U.S. time series: Total Policing and Cumulative Deindustrialization

($r2$= -.705, n= 32, p<.001)

the relationship between policing, inequality, exploitation, unionization, strikes and lockouts, and deindustrialization. This empirical assessment used existing macroeconomic statistics that were operationalized to suit our analysis and to conform to the core concepts being tested. The analysis confirmed the following five hypotheses:

> *Inequality will positively correlate to total policing;*
> *Surplus value will positively correlate to total policing;*
> *Unionization will inversely correlate to total policing;*
> *Unionization will positively correlate to strikes and lockouts;* and
> *Cumulative deindustrialization will positively correlate to total policing.*

The associations we have discovered are powerful and recurring both over time and across international borders where variations in legal contexts and institutional histories of policing are many and complex. The empirical verification of these basic tenets of a Marxian political economy of policing are significant for both the study of pacification as well as policing and security research more broadly.

Although aimed at the level of a larger political economy, the effect of increased policing, decreased unionization, and more inequality has profound institutional effects that structurally condition our everyday social relations. As Althusser once mused: if a police officer simply says: "Hey, you there!" in public, the individual who turns around "by this mere one-hundred-and-eighty-degree physical conversion... becomes a *subject*."[80] The multiplication of these institutionally mediated (interpellated) effects have significant implications on how we see the world, the power of ideology, and the rise to prominence of *security as hegemony*.[81] Thus, these empirical results may usefully inform a general social theory and resistance that increasingly identifies "policing" – broadly defined – as a core element of the global pacification of labour and the enforcement of capitalist relations.[82]

NOTES:

1. This paper is an aggregation of two previously published studies: George S. Rigakos & Aysegul Ergul, "Policing the Industrial Reserve Army: International Study" *Crime, Law, Social Change* Vol. 56, 2011, pp. 329-371; and George S. Rigakos & Aysegul Ergul, "The Pacification of the American Working Class: A Time-Series Analysis" *Socialist Studies* Vol. 9, No. 2, 2013, pp. 167-198.
2. Derek Thompson, "Income Inequality is Killing the Economy, Obama Says – Is he Wrong?" *The Atlantic*, April 3, 2012; http://www.theatlantic.com/business/archive/2012/04/income-inequality-is-killing-the-economy-obama-says-is-he-wrong/255407/.
3. Ariana Huffingtan, *Third World America: How Our Politicians Are Abandoning the Middle Class and Betraying the American Dream* (New York: Broadway Paperbacks, 2005).
4. Warren Buffet, "A Minimum Tax for the Wealthy" *New York Times*, November 25, 2012; http://www.nytimes.com/2012/11/26/opinion/buffett-a-minimum-tax-for-the-wealthy.html?_r=0.
5. See, Richard Wolff, "The Keynesian Revival: A Marxian Critique" in C. Fanelli, C. Hurl, P. Lefebvre, and G. Ozcan, eds., *Saving Global Capitalism: Interrogating Austerity and Working Class Responses to Crises* (Ottawa: Red Quills Books, 2011).
6. Karl Marx and Friedrich Engels, *The Communist Manifesto* (New York: Penguin Books, 1987)(orig.1850).
7. Buffet, "A Minimum Tax for the Wealthy."
8. See, Naomi Klein, *The Shock Doctrine: The Rise of Disaster Capitalism* (Toronto: Alfred A. Knopf, 2008); Adrian Smith, "Pacifying the 'Armies of Offshore Labour' in Canada" *Socialist Studies* Vol. 9, No. 2, 2013, pp. 78-93.
9. Lewis Uchitelle, "Job insecurity of workers is a big factor in Fed Policy" *New York Times*, February 27, 1997.
10. See, Wolff, "The Keynesian Revival: A Marxian Critique."
11. George S. Rigakos, "'To extend the scope of productive labour': Pacification as a police project" in M. Neocleous and G. S. Rigakos, eds., *Anti-security*, (Ottawa: Red Quill Books, 2011), pp. 57-83.
12. Mark Neocleous, *The Fabrication of Social Order: A Critical Theory of Police Power* (London: Pluto Press, 2000).
13. Mark Neocleous, "Security as Pacification" in *Anti-security*, pp. 23-56, p. 38.
14. Mark Neocleous and George S. Rigakos, "Anti-Security: A Declaration" in *Anti-security*, pp. 15-22.
15. George S. Rigakos, John L. McMullan, Joshua Johnson, and Gulden Ozcan, eds., *A General Police System: Political Economy and Security in the Age of Enlightenment* (Ottawa: Red Quill Books, 2009).
16. Rigakos, *Anti-Security*, pp. 70-71.
17. Neocleous, *The Fabrication of Social Order*, p. 20.
18. See, Rigakos et.al., *A General Police System*.
19. Morris Friedman, *The Pinkerton Labor Spy* (New York: Wilshire Book Co., 1907).
20. Frank Morn, *"The Eye that Never Sleeps": A History of the Pinkerton National Detective Agency* (Bloomington: Indiana University Press, 1982).
21. Bernard Hogg, "Public Reaction to Pinkertonism and the Labor Question" *Pennsylvania History* Vol. 11, July 1944, pp. 171-99.
22. S. R. Couch, "Selling and reclaiming state sovereignty: The case of the coal and iron police" *Insurgent Sociologist* Vol. 10, No. 1, 1981, pp. 85-91.

23. Across the globe, trade liberalization, and 'contracting out' policies under neo-liberalism have helped the security industry prosper. Security firms are now engaged in policing functions that only decades previous would have been viewed as a public responsibility: patrols of outdoor business districts and massive urban commercial and residential complexes; the guarding of penal institutions, nuclear facilities, seaports, airports, and even police stations. See, David. H. Bayley and Clifford Shearing, "The future of policing" *Law and Society Review* Vol. 30, 1996, pp. 585-606; Mark Button, *Security Officers and Policing: Powers, Culture and Control in the Governance of Private Space* (Aldershot: Ashgate, 2007); Les Johnston, *The Rebirth of Private Policing* (London: Routledge, 1992); Trevor Jones and Tim Newburn, *Private Security and Public Policing* (New York: Oxford/Clarendon, 1998); George S. Rigakos, *The New Parapolice: Risk Markets and Commodified Social Control* (Toronto: University of

Toronto Press, 2002); George S. Rigakos, *Nightclub: Bouncers, Risk and the Spectacle of Consumption*. Montreal: McGill-Queen's University Press, 2008); Clifford D. Shearing and Philip C. Stenning, "Reframing policing" in C. D. Shearing and P. C. Stenning, eds., *Private Policing* (Newbury Park: Sage, 1987), pp. 9-18; Allison Wakefield, Alison, *Selling Security: The Private Policing of Public Space* (Devon UK: Willan Publishing, 2003). Marxist analysts have long been aware of the role played by the public and private police in suppressing worker's organizations and their pivotal function for inculcating a wage-labour system at the height of mercantile capitalism. See, Sidney L. Harring and Lorraine McMullin, The Buffalo police 1872-1900: Labour unrest, political power and the creation of the police institution. *Crime and Social Justice* 5, 1975, pp. 5-14; Couch, "Selling and reclaiming state sovereignty," pp. 85-91; Robert Weiss, "The emergence and transformation of private detective industrial policing in the United States, 1850-1940" *Crime & Social Justice* 9, 1978, pp. 35-48; John L. McMullan, "Social surveillance and the rise of the 'police machine'" *Theoretical Criminology* 2, 1998, pp. 93-117; Neocleous, *The Fabrication of Social Order*; Rigakos et al., *A General Police System;* Robert Storch, Robert, "The plague of the blue locusts: Police reform and popular resistance in Northern England 1840-1857" *International Review of Social History* 20, 1975, pp. 61-90. In more recent years, they have pointed to the rise of private security in spaces of consumption, the emergence of aggressive forms of "parapolicing," and have theorized how the "security commodity" itself operates in a capitalist system. See, Volker Eick, "Preventive urban discipline: Rent-a-cops and neoliberal glocalization in Germany" *Social Justics* Vol. 33, 2006, pp. 1-19; Steven Spitzer and Andrew T. Scull, "Privatization and capitalist development: The case of the private police" *Social Problems* Vol. 25, 1977, pp. 18-29; Rigakos, *The New Parapolice*; Mark Neocleous, "Security, commodity, fetishism" *Critique* Vol. 35, 2007, pp. 339-355; George S. Rigakos, "Hyperpanoptics as commodity: The case of the parapolice" *Canadian Journal of Sociology* Vol. 23, 1999, pp. 381-409; Geroge S. Rigakos, "Beyond Public-Private: Toward a New Typology of Policing" in D. Cooley, ed., *Re-Imagining Policing in Canada* (Toronto: University of Toronto Press, 2005), pp. 260-319; Stephen Spitzer, "Security and control in capitalist societies: The fetishism of security and the secret thereof" in J. Lowman, R. J. Menzies, and T. S. Palys, eds., *Transcarceration: Essays in the Sociology of Social Control, Cambridge Studies in Criminology* (Aldershot: Gower, 1987), pp. 43-58. In the end, both public and private police originate from the same liberal capitalist principles that are concerned about infinite market expansion and capital accumulation.
24 Neocleous and Rigakos, *Anti-Security*, pp. 15-16.
25 Adam Smith, *The Wealth of Nations* (New York: Random House, 1937) (orig. 1776), p. 18.
26 Karl Marx, *Wage-Labor and Capital* (New York: International Publishers, 1977) (orig. 1933), p. 33.
27 See, Rigakos et al., *A General Police System*.
28 Karl Marx, *Theories of Surplus-Value, I* (London: Lawrence and Wishart, 1972).
29 Antonio Gramsci, "Soviets in Italy" *New Left Review* Vol. 51, 1968, pp. 28–50, p. 34.
30 Rosa Luxemburg, Rosa, "Social Reform or Revolution" in D. Howard, ed., in *Selected Political Writings* (London: Monthly Review Press, 1971), pp. 52-135, p. 68.
31 Leon Trotsky, "The Trade Unions in Britain" in *Collected Writings and Speeches on Britain Vol. III* (London: New Park Publications, 1974), p. 43.
32 Leon Trotsky, *Leon Trotsky on the Trade Unions* (London: Merit Publishers, 1969), p. 54.
33 Karl Marx, "On Wages, Hours, and the Trade-Union Struggle" in K. Lapides, ed., *Marx and Engels on the Trade Unions* (New York: Praeger, 1987), pp. 90-93.
34 See, Neocleous, *The Fabrication of Social Order*.
35 See, Couch, "Selling and reclaiming state sovereignty," pp. 85-91; Weiss, "The emergence and transformation of private detective industrial policing in the United States, 1850-1940," pp. 35-48.
36 David Priestland, *The Red Flag: A History of Communism* (New York: Grove Press, 2011).
37 For our international comparison it was sometimes necessary to transpose annual data from other years (usually a one-year deviation) in order to create a complete dataset. Indeed, this also explains why our snapshot year is 2004. Worldwide surveys and national accounting statistics from emerging and post-Soviet economies make obtaining up-to-date statistics almost impossible for non-EU and non-NAFTA nations. Current police employment statistics are also sometimes considered a national security matter in certain countries (e.g. Turkey and Greece).

38 These sources are: (1) the European Sourcebook of Criminal Justice Statistics (2006); (2) the Panoramic Overview of the Private Security Industry in the 25 Member States of the European Union (2004) by the European Commission for the Confederation of European Security Services (CoESS); and (3) the report on SAWL and Private Security Companies in South Eastern Europe (2005) by the South Eastern Europe Clearinghouse for the Control of Small Arms and Light Weapons (SEESAC).
39 More specifically: (1) a combination of census and FBI *Law Enforcement Employment Bulletins* and (2) Occupational Employment Statistics by the Bureau of Labour Statistics (BLS).
40 The WVS was conducted in a number of successive "waves" including 1981, 1990-1991, 1995-1996, and 1999-2001. In this paper, only the results of the most recent survey are employed. Another WVS wave took place from 2003-2005, however only Kyrgyzstan, Hong Kong and Morocco were asked the unionization question.
41 The data retrieved from OECD. Stat Extracts includes: Australia, Austria, Belgium, Canada, Czech Republic, Denmark, Finland, France, Germany, Greece, Hungary, Ireland, Italy, Luxemburg, Netherlands, Norway, Poland, Portugal, Slovakia, Spain, Sweden, Switzerland, Turkey, UK, Japan, Republic of Korea, USA.
42 The data retrieved from eironline includes: Bulgaria, Cyprus, Estonia, Latvia, Romania, and Slovenia.
43 Gerald Mayer, *Union Membership Trends in the United States*, Congressional Report Series, Washington, 2004.
44 Barry T. Hirsh and David Macpherson, *Union Density Estimates by State, 1964-2011*, November 6, 2012; http://www.unionstats.com
45 Developed by the Italian statistician Corrado Gini (1921), the coefficient is the ratio of the area under a line of equality where one axis is the cumulative share of income and the other axis is the cumulative share of people from the lowest to the highest. The coefficient produces a range from zero to one and is often multiplied by 100. The higher the Gini coefficient, the higher the rate of income inequality. If G=0, then everyone receives the same percentage of income – a completely egalitarian society. If G=100 then one person has all the income – a completely unequal society. See, Corrado Gini, "Measurement of inequality of incomes" *The Economic Journal* Vol. 31, 1921, pp. 124-126.
46 See, Alice H. Amsden, "An International Comparison of the Rate of Surplus Value in Manufacturing Industries" *Cambridge Journal of Economics* Vol. 5, 1981, p. 229–249; Carl J. Cuneo, "Reconfirming Marx's Rate of Surplus Value" *Canadian Review of Sociology and Anthropology* Vol. 21, 1984, pp. 98-104; Carl J. Cuneo, "Class Struggle and Measurement of the Rate of Surplus Value" *Canadian Review of Sociology and Anthropology* Vol. 19, 1982, pp. 377-425; Carl J. Cuneo, "Class Exploitation in Canada" *Canadian Review of Sociology and Anthropology* Vol. 15, 1978, pp. 284-300; Michael J. Lynch, "The extraction of surplus value, crime and punishment" *Contemporary Crises* Vol. 12, 1988, pp. 329-344; Michael J. Lynch, "Quantitative Analysis and Marxist Criminology: Some Solutions to the Dilemma of Marxist Criminology" *Crime and Social Justice* Vol. 29, 1987, pp. 110-127; Fred Moseley, "The rate of Surplus Value in the Postwar US Economy: A Critique of Weisskopf's Estimates" *Cambridge Journal of Economics* Vol. 9, 1985, pp. 57-79; Thomas E. Weisskopf, "The Rate of Surplus Value in the Postwar U.S. Economy: A Response to Moseley's Critique" *Cambridge Journal of Economics* Vol. 9, 1985, pp. 81-84; Edward N. Wolff, "The Rate of Surplus Value in Puerto Rico" *Journal of Political Economy* Vol. 83, 1975, pp. 935-949; Edward N. Wolff, "The Rate of Surplus Value, the Organic Composition, and the general rate of Profit in the U.S. Economy" *American Economic Association* Vol. 69, 1979, pp. 329-341.
47 See, Cuneo, "Class Exploitation in Canada," pp. 284-300; Lynch, "The extraction of surplus value, crime and punishment," pp. 329-344; Michael J. Lynch, W. Byron Groves, and Alan Lizotte, "The rate of surplus value and crime. A theoretical and empirical examination of Marxian economic theory and criminology" *Crime, Law and Social Change* Vol. 21, 1994, pp. 15-48.

48 The data required for the calculation of surplus value consists of: the number of employed production workers, manufacturing value added, and annual earnings of manufacturing workers (production workers). The measures that comprise our formula are derived from: (1) a custom data retrieval of *gross value-added* from the World Bank (Development Indicator); and (2) a secondary calculation of *annual earnings of manufacturing workers* based on data derived from the ILO (LABORSTA Internet) utilizing total employment by economic activity (i.e. manufacturing), wages in manufacturing, and hours of work in manufacturing. These calculations were then made comparable by converting all foreign currencies into U.S. dollars. We used a four-year average leading up to and including our snapshot year of 2004.
49 These are: (1) Current Employment Statistics on Employment, Hours and Earnings, produced by BLS; (2) Gross Domestic Products Accounts by Industry, created by the U.S. Bureau of Economic Analysis; and (3) Annual Survey of Manufacturers and American Fact Finder, prepared by the U.S. Census Bureau.
50 See, David Brauner, *Factors Underlying the Decline in Manufacturing Employment Since 2000.* Economic and Budget Issue Brief, 2008, Congressional Budget Office.
51 See, David Held & Kaya, Ayse, "Introduction" in David Held and Ayse Kaya, eds., *Global Inequality: patterns and explanations* (Cambridge: Polity Press, 2007), pp. 1-25.
52 See, Anders Åslund, *How Capitalism was Built: The Transformation of Central and Eastern Europe, Russia and Asia* (New York: Cambridge University Press, 2007).
53 See, Matthew Thomas Clement, "Is rising global inequality a myth?" in *Monthly Review*, August 19, 2008; http://mrzine.monthlyreview.org/2008/clement190808; Y. Dikhanov and M. Ward, *Evolution of the Global Distribution of Income in 1970-99*, Initiative for Policy Dialogue, Columbia University, 2003; Jon Kenneth Galbraith, "Responses: Is inequality decreasing?: By numbers" *Foreign Affairs*, July/August 2002, pp. 178-179; Branko Milanoviç, "Globalization and Inequality" in D. Held and A. Kaya, eds., *Global Inequality: Patterns and Explanations* (Cambridge: Polity Press, 2007), pp. 26-49; J. W. Pitts, "Responses: Is inequality decreasing?: Inequality is no myth" *Foreign Affairs*, July/August 2002, pp. 179-180; Bob Sutcliffe, "The Unequalled and Unequal Twentieth Century," in *Global Inequality: Patterns and Explanations*, pp. 50-72; R. H. Wade, "Is globalization reducing poverty and inequality?" *World Development* Vol. 2, 2004, pp. 1-23.
54 See, Milanoviç, *Global Inequality*, pp. 26-49; Wade, "Is globalization reducing poverty and inequality?," pp. 1-23. Neoliberal scholars have argued that global income inequality between nations has been declining since around the 1980s. See, Martin Wolf, "The Big Lie of Global Inequality" *The Financial Times*, January 24, 2000; David Dollar, & Aart Kraay, "Spreading the Wealth" *Foreign Affairs*, January/February 2002, pp. 120-133; Glenn Firebaugh, *The New Geography of Global Income Inequality* (Cambridge: Harvard University Press, 2003); David Dollar, "Globalization, Poverty, and Inequality since 1980" *The World Bank Research Observer* Vol. 20, 2005, pp. 145-175; François Nielsen, "Income Inequality in the Global Economy: the myth of rising world inequality" *Harvard College Economics Review* Vol. 1, 2007, pp. 23-26.
55 John Kenneth Galbraith, "A perfect crime: Inequality in the age of globalization" *Daedalus* Vol. 131, 2002, pp. 11-15; John Kenneth Galbraith, "Global inequality and global macroeconomics" in *Global Inequality: Patterns and Explanations*, pp. 1-25; Wade, "Is globalization reducing poverty and inequality?," pp. 1-23.
56 See, Human Development Reports, *Overview of the Human Development Report*, 1999; http://hdr.undp.org/en/reports/global/hdr1999/; Human Development Reports, *Summary of the Human development Report 2005*; http://hdr.undp.org/en/media/hdr05_summary.pdf
57 See, Kevin Watkins, *Human Development Report 2005*, United Nations Development Programme, New York.
58 See, James Davies, Susanna Sandström, Anthony Shorrocks, and Edward Wolff, *The World Distribution of Household Wealth*, World Institute for Development Economics Research of the United Nations (UNU-WIDER), New York, 2005.
59 See, Andrea Brandolini and Timothy M. Smeeding, "Patterns of economic inequality in western democracies: some facts on levels and trends" *Political Science and Politics* Vol. 39, No. 1, 2006, pp. 21-26; Timothy M. Smeeding, "Public Policy and Economic Inequality: The United States in Comparative Perspective" *Social Science Quarterly* Vol. 86, 2005, pp. 955-983; Martina Morris and Bruce Western, "Inequality in earnings at the close of the twentieth century" *Annual Review of Sociology* Vol. 25, 1999, pp. 623-657.

60 See, U.S. Census Bureau, Gini Ratios for Households by Race and Hispanic Origin of Householders:1967-2010,December 19, 2011; http://www.census.gov/hhes/www/income/data/historical/inequality/index.html; Linda Levine, *U.S. Income Distribution and Mobility: Trends and International Comparisons*, Congressional Research Service, Washington, 2012.
61 See, Neocleous, *The Fabrication of Social Order.*
62 See, Åslund, *How Capitalism was Built.*
63 Lumpenproletariat is also often erroneously conflated with "surplus population" and "industrial reserve army." See also Bovenkirk's thorough and illuminating critique of Marx and Engel's use of the term "lumpenproletariat" as a rhetorical device: Frank Bovenkerk, "The rehabilitation of the rabble: How and why Marx and Engels wrongly depicted the lumpenproletariat as a reactionary force" *The Netherlands Journal of Sociology/Sociologia Neerlandica* Vol. 20, 1984, pp. 13-41.
64 Karl Marx, *Capital, 1.*, B. Fowkes, trans. (New York: Penguin, 1976) (orig. 1867), p. 1044.
65 Scholars have debated the relative merits of Marx's inconsistent notions of productive and unproductive labour for some time but what is largely agreed upon is that Marx's treatment of productive labour while initially expansive in critique and contradistinction to Smith becomes increasingly narrow with further clarification and focuses more and more on the relative position of workers to direct production. See, Ian Gough, "Marx's theory of productuve and unproductive labor" *New Left Review* Vol. 76, 1972, pp. 47-72; C. A. Hertig, "Developing productive realtionships with private security" *FBI Law Enforcement Bulletin* Vol. 55, 1986, pp. 19-22; David Houston, "Productive and unproductive labor: Rest in peace" *Review of Radical Political Economics* vol. 29, 1997, pp. 131-147; E.K. Hunt, "The categories of productive and unproductive labor in Marxist economic theory" *Science and Society* Vol. 43, 1979, pp. 303-325; Fyodor I. Kushnirsky and William J. Stull, "Productive and Unproductive Labour: Smith, Marx, and the Soviets." in D. A. Walker, ed., *Perspectives on the History of Economic Thought, Selected Papers from the History of Economics Society Conference 1987* (Aldershot: Gower, 1989); David Leadbeater, "The consistency of Marx's categories of productive and unproductive labour" *History of Political Economy* Vol 17, 1985, pp. 591-618; Ernest Mandel, *Introduction to Capital Vol.1* (London: Penguin, 1976), pp. 11- 86; Simon Mohun, "Productive and unproductive labor in the Labor Theory of Value" *Review of Radical Political Economics* Vol. 28, 1996, pp. 30-54; Guglielmo Carchedi, *On the Economic Identification of Social Classes* (London: Routledge & Kegan Paul, 1977).
66 See, B. Ebbinghaus and J. Visser, "When Institutions Matter — Union Growth and Decline in Western Europe, 1950-1995" *European Sociological Review* Vol. 15, 1999, pp. 135-158; Herbert Kohl, *Where Do Trade-Unions Stand Today in Eastern Europe? Stock-Taking after EU Enlargement*, Friedrich-Ebert-Stiftung International Trade-Union Cooperation, 2008; Jelle Visser, "Union membership statistics" *Monthly Labour Review* January 2006, pp. 38-49; David G. Blachflower, *A Cross-Country Study of Union-Membership*, Institution for the Study of Labour (IZA), Universität Bonn, 2006; Richard P. Chaykowski, and George A. Slotsve, "Earnings inequality and unions in canada" *British Journal of Industrial Relations* Vol. 40, No. 3, 2002, pp. 493-519; Steven Greenhouse, "Sharp Decline in Union Membership in '06" *The New York Times*, January 26, 2007; Andrew Leigh, "The Decline of an Institution" *Australian Financial Review*, March 7, 2005.
67 See, Ebbinghaus and Visser, "When Institutions Matter," pp. 135-158; H. Lesch, "Trade Union Density in International Comparison" *CESifo Forum* Vol. 4, 2004, pp. 12-18; M. Wallerstein and B. Western, "Unions in Decline? What has Changed and Why" *Annual Review of Political Science* Vol, 3, 2000, pp. 355-377.
68 See, Bernd Fitzenberger, Karsten Kohn, and Qingwei Wang, *The Erosion of Union Membership in Germany: Determinants, Densities, Decompositions*, Institution for the Study of Labour (IZA), Universität Bonn, 2006.
69 See, Leigh, "The Decline of an Institution."
70 See, Ebbinghaus and Visser, "When Institutions Matter," pp. 135-158.
71 See, B. Ebbinghaus, "Trade unions' changing role: Membership erosion, organizational reform, and social partnership in Europe" *Industrial and Labor Relations Review* Vol. 33, 2002, pp. 465-483; H. C. Katz, "The decentralization of collective bargaining: A literature review and comparative analysis" *Industrial and Labor Relations Review* Vol. 47, 1993, pp. 1-22.
72 See, J. Pencavel, *The Surprising Retreat of Union Britain*, Institution for the Study of Labour (IZA), Universität Bonn, 2003; Wallerstein and Western, "Unions in Decline?".

73 R. Freeman and J. Pelletier, "The impact of industrial relations legislation in British union density" *British Journal of Industrial Relations* Vol. 28, 1990, pp. 141-164; Pencavel, *The Surprising Retreat of Union Britain*.
74 See, P. Böckerman and R. Uusilato, *Union Membership and the Erosion of the Ghent System: Lessons from Finland*, Labour Institute for Economic Research, 2005; Lesch, "Trade Union Density".
75 See, Vadim Volkov, "The political economy of protection rackets in the past and the present" *Social Research* Vol. 67, 2000, pp. 709-744.
76 See, Till Von Wachter, "Employment and Productivity Growth in Service and Manufacturing Sectors in France, Germany and the US" *European Central Bank Working Paper* No. 50, 2001, pp. 1-59; Anita Wölfl, "Productivity Growth in Service Industries: An Assessment of Recent Patterns and the Role of Measurement" *OECD Science, Technology and Industry Working Papers* 2003/7, pp. 1-67; Eric O'N. Fisher, "Why are We Losing Manufacturing Jobs?" *The Federal Reserve Bank of Cleveland*, July 2004, pp. 1-4; Eric O'N. Fisher & Peter C. Rubert, "The Decline of Manufacturing Sector in the United States" *Working Paper* 2004, The Ohio State University, pp. 1-34; Anita Wölfl, "The Service Economy in OECD Countries" *OECD Science, Technology and Industry Working Papers* 2005/3, pp. 1-82; Dirk Pilat, Agnés Cimper, Karsten B. Olsen & Colin Webb, "The changing Nature of Manufacturing in OECD Countries" *OECD Science, Technology and Industry Working Papers* 2006/9, pp. 1-38; Frans van der Zee & Felix Brandes, "The Manufacturing Futures for Europe: A Survey of the Literature" *TNO* 2006, pp. 1-46; Felix Brandes, "The Future of Manufacturing in Europe: A Survey of the Literature and a Modelling Approach" *The European Foresight Monitoring Network* (EFMN), Brussels, 2008.
77 See, Robert Zieger, and Gilbert J. Gall, *American Workers, American Unions: the twentieth century* (Baltimore: The John Hopkins University Press, 2002).
78 See, Megan M. Baker, "Manufacturing Employment Hard Hit During the 2007-2009 Recession" *Monthly Labour Review*, April 2011, pp. 28-33.
79 See, Brauner, *Factors Underlying the Decline in Manufacturing Employment Since 2000.*
80 Louis Althusser, "Ideology and Ideological State Apparatuses" in Ben Brewster, trans., *Lenin and Philosophy and Other Essays* (New York: Monthly Review Press, 1971), p. 174.
81 George S. Rigakos and Martin V. Manolov, "Anti-Security: Q & A Interview of George S. Rigakos" *The Annual Review of Interdisciplinary Justice Research* 3, 2013, pp. 9-26, p. 16-19.
82 Neocleous and Rigakos, *Anti-Security*.

Unmanning the Police Manhunt
Vertical Security as Pacification

Tyler Wall

Why, oh why must you swoop through the hood
Like everybody from the hood is up to no good
Run, run, run from the ghetto bird Run.

Ice Cube, "Ghetto Bird" (1993)

In the name of "security," battlefronts bleed into home fronts as military technologies charged with the pacification of foreign others "outside" national space are tasked with the pacification of others on the "inside." This is perhaps most evident with the emergence of Unmanned Aerial Vehicles (UAVs), or aerial surveillance drones, as they migrate from the securityscapes in Iraq, Afghanistan, and Pakistan to the United States "homeland." Known for their powerful surveillance cameras, thermal imaging, hovering capabilities, aerial flexibility and, depending on the model, destructive missile strike capabilities, drones have emerged as a contemporary icon of the cutting edge of air power. US Secretary of Defense Leon Panetta once stated that drones are "the only game in town" in terms of combatting "terrorism"[1]—a logic embraced by an Obama administration seemingly undeterred from accumulating civilian deaths while expanding and ramping up drone attacks premised on a secretive "kill list" of "suspected terrorists," including US citizens.[2] Clearly, aerial drones are not merely a game, as Panetta would have it, but indeed a bloody business mobilized by the imperatives of security and accumulation. Yet the drone market is not confined to foreign theaters, as the US security state and security industries are increasingly imagining drones as "dual-use" scopic technologies that can readily be deployed across many spatial contexts removed from foreign policy, at least on the surface.[3] One such context is the policing of domestic order, especially what is commonly but problematically referred to as routine "law enforcement" or "crime fighting." That is, military aerial drones are now being "repurposed" as domestic security technologies.

As the case of aerial drones demonstrates the contemporary politics of security is routinely measured through a "technological fix," most commonly through a visual prosthetics pregnant with the possibility of violence.[4] This fixation on seeing, knowing, and ordering through optical enhancement can also be seen with the ubiquitous information and biometric technology such as body scans, facial recognition systems, smart cards, national ID cards, cell-phone tracking devices, geospatial satellite-tracking devices, Closed-Circuit Television (CCTV), and a plethora of other technologies aimed at collecting "intelligence." All of these coercive looking technologies convert information into "intelligence" through the mediating capacities of screens, databases, and networks that function by abstracting bodies from their local contexts to facilitate various interventions.[5] Just like UAV systems, all of the above technologies have been and are currently deployed in both counter-insurgency and domestic policing operations – suggesting that these technologies never solely belong to the domestic order, but to the order of security and pacification. This order is rooted in the "boomerang effect," whereby control technologies deployed abroad in colonial and military campaigns "boomerang" back to the metropole to be deployed against "homefront" populations, and most often against those populations deemed "threatening" and "disorderly" such as the poor, racial minorities, and social movements.[6]

The case of police drones speaks directly to the importation of actual military and colonial architectures into the routine spaces of the "homeland," disclosing insidious entwinements of war and police, metropole and colony, accumulation and securitization. Yet the pervasive trafficking of technologies between military and police are often met with a persistent denial, namely, the normalization of a pervasive assumption that imagines "colonial frontiers and Western 'homelands' as fundamentally separate domains" rather than seeing these spaces as "fuse(d) together into a seamless whole."[7] But, as I attempt to demonstrate, it would simplistic and misleading to suggest that the pacification of foreign populations and securing of global markets, to which military drones have played an important part of late,[8] is somehow removed from the pacification of domestic territory and securing of markets on the "inside." Although at the time of this writing unmanned vertical policing is not yet widespread, making the analysis admittedly speculative, my purpose here is to demonstrate the union of war power abroad and police power at home. Police drones then must be understood as continuous with, and in no way detached or dissimilar from, contemporary US pacification projects in Iraq, Afghanistan, and Pakistan. The article therefore unpacks how police UAVs, like the military drone, are bounded by the logic of security and the practice of pacification as these vertical tracking technologies are tasked with the hunting of human prey. In this sense, police drones underline the unmanning of the police manhunt, that foundational practice of police power where the "reserve army of labour" is quite literally hunted and captured.

THE RISE OF DRONE PATROLS

While unmanned military commodities have been profitable for the US security industry, the uncertainty around future budget cuts and wars drives the industry to hunt for new "internal" drone markets and the removal of "obstacles" to capital accumulation. This is one face of capital's perpetual hunt for new markets.[9] Faced with the fear of future budget cuts and pending wars, the US security state and partnered security industries are persistently manufacturing "adjacent markets," or any civilian market where military technologies can be peddled.[10] A defense executive has stated that the industry goal for military ISR [Intelligence, Surveillance, and Reconnaissance] technologies is "to push it down to the state and local governments to see if there is a mission to support."[11] Importantly, the "mission support" mentioned by the above executive is the "public safety market." As a different defense executive states: "a number of our influential products have dual-use capability to locations and missions adjacent to our primary overseas ISR mission. One such example is local law enforcement, emergency first responders and border protection."[12] The military drone is at the forefront of the so-called green-to-blue pipeline, or the movement from military to domestic security applications.

Prior to 2012 there had been one major obstacle to domesticating drones, namely, Federal Aviation Administration (FAA) regulations blocking widespread access to national airspace by both public and private institutions. In February 2012 this obstacle, if not completely demolished, was reworked into a much less significant impediment with passage of H.R. 658, a law requiring the FAA to expedite the process of handing out Certificate of Authorizations (COAs) to government agencies such as the police and border patrol and also private enterprises so that they can operate micro-drones. It has been estimated that by 2018, there could be 30,000 drones flying in US skies—a mixture of military, public safety, and private drones.[13] The passing of the bill was largely due to sustained pressure by drone stakeholders, primarily Congress's Unmanned Systems Caucus, the Association of Unmanned Vehicles International (AUVSI) and its corporate members, Department of Homeland Security (DHS), various lawmakers, and domestic policing agencies. These stakeholders argued that the lack of access to US airspace was a hindrance to both capital accumulation and much needed security measures. As a spokesperson for the AUVSI has stated, "The potential civil market for these systems could dwarf the military market in the coming years if we can get access to the airspace." Michael Huerta, an FAA administrator, has stated: "What we're hearing from the Congress and the industry is, 'This technology is evolving quickly and we don't want the FAA to be too cautious so as to hold up technological innovation.'"[14]

Unsurprisingly, "public safety" agencies across the US have embraced this move to "re-purpose" and "re-deploy" military-style UAVs, specifically micro-drones weighing from 4 to 25 pounds and from 2 to 8 feet in length. A Texas official has stated, "Public-safety agencies are beginning to see this as an invaluable tool for them, just as the car was an improvement over the horse and the single-shot pistol was improved upon by the six-shooter."[15] To police drone enthusiasts, UAV systems evoke a "technological sublime,"[16] or a certain reverence, awe, and arousal concerning great engineering feats and technologies. In this case, drones are a technological sublime that points to the dream of securing the insecurity of domestic order. Outfitted with potent cameras and potentially night vision, facial recognition, thermal imaging, and even lethal and non-lethal weaponry, drones are said to be a dreamlike, "silver-bullet" scopic commodity animating the fantasy of security. Police micro-UAVs have been imagined for a plethora of circumstances: natural disaster assistance, search and rescue, special events and other large gatherings such as protests, traffic congestion and enforcement, high speed pursuits, locating fleeing/hiding suspects, hostage rescue and barricaded subjects, drug interdiction, and surveillance/intelligence operations. Indeed, the police applications of this appear endless, with innovation a likely outcome of their adoption in everyday police practices. As one spokesperson for a local government that purchased a drone remarked, "As we get into this we'll be able to find more uses for it."[17]

Perhaps the most well-known case of domestic UAVs is the implementation of drones by the Department of Homeland Security (DHS) in the aerial monitoring of US border regions with Mexico and Canada.[18] Currently, the CBP has 9 drones with plans for more in the future.[19] But drones are also emerging beyond the seams of US borderscapes as increasing numbers of US police departments are seeking military-style aerial drones as key domestic policing technologies. To list only a few examples, drones have been acquired by FAA authorizations or have been applied for by policing agencies in Seattle, Colorado, Texas, Maryland, California, North Dakota, Florida, South Carolina, Alabama, Utah, Idaho, and Arkansas. For instance, Miami-Dade police received a grant from the Department of Justice to acquire two Honeywell T-Hawk drones, at $50,000 each, that can fly and hover at altitudes up to 9000 feet. The local government of Canyon County, Idaho purchased a Draganflyer X6 with DHS grants.[20] Like other similar drones, the Draganflyer X6 can stream video to officers on the ground and also comes equipped with thermal imaging technology. The Texas Department of Public Safety (DPS) has four Wasp III drones that reportedly are available on a case-by-case basis to any policing agency in the state.[21] In October of 2011 the Sheriff's Department in Montgomery County, Texas, also with assistance from DHS grants, unveiled a 7-foot long drone called the "Shadowhawk." This particular drone, from Texas-based Vanguard Industries, is equipped with cameras and heat sensor and night vision technology and the platform can be armed with "non-lethal" and "lethal" weaponry. As of May 2012, it was reported that the Shadowhawk had yet to be deployed, but officials stated that they were waiting for the "right incident" to "present itself."[22] It is not an understatement to say that both the idea and the reality of police drones have become normalized in policing circles. As one New York Police Department (NYPD) spokesperson puts it, drones just "aren't that exotic anymore."[23]

Despite all of these developments, the opening of the police drone market has been met with critiques from liberals and conservatives alike, ranging from concerns about safety concerns such as mid-air collisions and loss of signal scenarios, even though the issue of privacy, unsurprisingly and problematically, has dominated popular critiques.[24] For the security industry these issues are to be solved through "public relations." Speaking at a Counter Terror Expo, a government official stated that "We have a very tall challenge to change public perception. Otherwise, we'll be stopped cold in our tracks if we don't do this thoughtfully. We have to bring the public along every step of the way" so that they realize "we will not be watching backyards."[25] Indeed, going so far as to hire a public relations firm to "bombard the American public with positive images and messages,"[26] the AUVSI has admitted that one of the big challenges for the emergence of domestic UAVs is winning "hearts and minds." An AUVSI spokesperson has stated that "We're going to do a much better job of educating people about unmanned aviation, the good and the bad. We're working on drafting the right message and how to get it out there. You have to keep repeating the good words. People who don't know what they're talking about say these are spy planes or killer drones. They're not."[27] However, there are more commonalities between military drones and police drones than this spokesperson suggests. The majority of military UAVs are primarily equipped for aerial surveillance and intelligence-gathering, and are not equipped with lethal systems and are not nearly the size of the Predator and Reaper "hunter-killer" drones that have received most attention. Indeed, one suggested solution to successfully normalize drones in national airspace is to cease calling the technology "drones," but rather "remotely piloted vehicles"[28] because the word "drone" is so associated with targeted assassinations, kill lists, and dead civilians.

Clearly, the rise of police drones reveals a bundle of issues concerning technologies of violence, questions of security, and the powers of marketing. How might we understand the police drone, without falling back on liberal worries about 'safety concerns' or loss of privacy? How might we situate the drone within the wider frame of the critique of security and the logic of pacification?

SECURITY FETISHISM AND INSECURITY AS OPPORTUNITY

Animated by the fetish of security,[29] the rise of US police drones exemplifies how logics of (in)security circulate and proliferate, and in so doing, create new configurations of state power and accumulation. Although police officials justify drones by claiming they are cheaper than helicopters and better protect officers from "harm's way," discourses of security remain the most forceful argument, as police officials routinely exclaim that drones offer an extra layer of "public safety." Prior to the passing of H.R. 658, New York Democrat Charles Schumer stated that the domestication of UAVs is ultimately a matter of "national security":

> The FAA has been very hesitant to give authorization to these UAVs due to limited air space and restrictions that they have. I certainly can appreciate those concerns; but when we're talking about Customs and Border Protection or the FBI, what have you, we are talking about missions of national security. And certainly there's nothing more important than that.[30]

Unmanned police power then can firmly be situated in what Feldman dubs "securocratic wars of public safety" where national security and public safety concerns converge and become inseparable.[31] As "the supreme concept of bourgeois society," as Marx once put it,[32] and a "general economy of power" as Foucault suggested,[33] security exercises an insidious mutability and malleability, and both writers also recognized the securing of insecurity as always unfinished and perpetual. Consequently, the "war on terrorism" slides into those other perpetual security projects, such as the "war on crime" and "war on drugs," while homeland security, public safety, and national security become interchangeable—hence the normalized overlapping techniques of military and police power in which drones are but one example.

Both imperial and domestic police UAVs are first and foremost security commodities invested in and bounded by the prerogatives of security and accumulation, accentuating how security becomes commodified in neoliberal "risk markets."[34] If the commodity form is said to address or alleviate some human need, and the security commodity is specifically that of insecurity, then the police drone addresses the local security state's need or desire or dream of pacifying territory and populations.[35] Police agencies turn to security industries in order to better enhance their security objectives while security industries aggressively market military products to "public safety" institutions to secure accumulation. That is, the emergence of police drones speak to the ways the security state and security industries are virtually indistinguishable, as attested to above with the entanglement of the AUVSI, DHS, Congressional Unmanned Caucus, policing agencies, and the FAA's relinquishing of the control of airspace due to the intertwined imperatives of security and accumulation.

This entails not so much the retreat or "hollowing-out" of the state or a privatization of the state, but a security industry intimately intertwined with the state.[36] The state appears as catering to an increasingly powerful security industry, with local police forces the main client. And yet there is no clear separation between the security provided by the state and the market of commodity circulation. Much like the "child protection industry,"[37] drone industries, and security corporations in general, do not produce and market security commodities such as UAVs because they are particularly interested in surveillance and security per se, but first and foremost because they are interested in accumulation.[38] Yet they also recognize that to secure accumulation, a healthy security state must be forged and nourished.

As attempts to domesticate drones suggest, the logic of security presupposes a social order or even local context that is haunted by the spectre of insecurity. That is, police drone stakeholders are reliant on the presence of what could be thought of as "opportunities of insecurity" that are often tragic, transgressive, or perceived as "disorderly" to help justify the continual reproduction, circulation, and intensification of the security-accumulation assemblage. Here it is useful to think of this in terms of the "disaster capitalism" outlined by Naomi Klein, in which human and environmental devastation is seen as an occasion for state power and capitalist accumulation to expand. As Klein shows, "homeland security" itself is largely an economy where unchecked police powers and unchecked capitalism insidiously converged after "9/11."[39] This point on disaster capitalism in reference to police drones is poignantly demonstrated by the following graphic by Lucintel, a market research group:

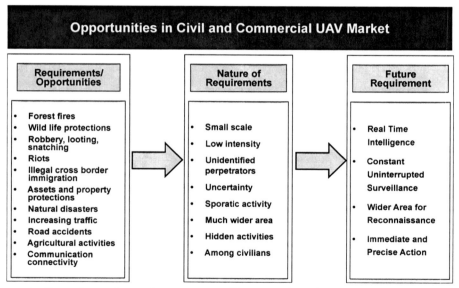

As the column above demonstrates, diverse and random events, most of them tragic or transgressive in that lives have been injured and/or order breached, are framed as not upsetting or disconcerting in terms of human suffering or harm, but as the necessary conditions for "growth" of the UAV civil market. Reworked as "requirements" and "opportunities," these various events ranging from forest fires to automobile accidents, from criminalized activity to an influx in foot or vehicular traffic, are understood as the required conditions of possibility for the domestic drone market. And as the middle column shows, these opportunities to capitalize on tragedy and transgression are structured by the uncertain and unpredictable, the ambiguous and illegible, as they play out on a local scale. In addition, the column on the right points to how the intertwined logics of security and insecurity, and order and disorder, animate the whole process, as the pursuit of security is understood as perpetual but never achievable. The spectre of blindness and ignorance haunts or animates security to develop and deploy greater capabilities to "see" and "order."

Police drones are but one useful case study demonstrating how the security state and security industries are reliant upon, and actively propagate and mobilize fear, suspicion, and anxiety around "public safety" issues to simultaneously justify hardened security measures and secure accumulation. Of course, this is nothing new per se, but it is just one example of the ways in which demands for security are at once demands for accumulation.[40] Indeed, security capitalizes on devastation and insecurity by converting them into a plethora of opportunities for state power, social order and capitalist accumulation to be bolstered and reproduced.[41] Thus Cowen and Siciliano speak of a "securitized social reproduction" whereby individual bodies, groups, asymmetrical social relations, and the larger order of capitalist accumulation are secured through classed, gendered, and racialized practices of targeted policing. This process, I suggest, needs to be understood as central to the process of pacification. The remainder of this article thinks through the cultural and political dynamics configuring the domestic deployment of police drones in this light. I suggest that aerial police drones are nothing less than a human hunting technology in the service of pacification.

PACIFICATION, SCOPIC VERTICALITY, AND THE MANHUNT

Police drones are often described as part of a long history of police "militarization" where both martial vocabularies and military practices and technologies[42] abound within modern policing's "technostructure."[43] Yet as tempting as the militarization discourse is we must resist seeing the boomeranging of military drones to the policing of domestic populations as solely an issue where martial logics and technologies straightforwardly convert police power into war power. The militarization discourse often lends itself to a problematic "blurring" trope where the military invades and corrupts, "militarizes," the previously noble police profession. As Guillermina Seri argues, "There is a distinct police role in facilitating authoritarianism and state violence,"[44] and this is precisely what is missed in discussions of the militarization of the police. The police and military might operate with "different notions of risk" such as arrest/prosecution as opposed to simply kill,[45] yet insisting on this divide obscures the fact that war and police have long been sutured together in the name of security. That is, the boomeranging of military UAVs is but one contemporary example of how war power and police power have long been allied, and it is the logic of security and the practice of pacification that animates both.[46]

Most often pacification is evoked in regards to military strategy and tactics, primarily in reference to counter-insurgency efforts to "win hearts and minds" in the US/Vietnam war—although it has a much longer colonial history.[47] But as others have pointed out, the "external pacification" of distant territories and peoples has historically developed alongside the "internal pacification" of domestic territories and populations,[48] the former primarily being consigned to the military whereas the latter a policing project. In his ethnography of the LAPD, Herbert identifies police surveillance as one important means by which the police routinely aim for "internal pacification."[49] In this regard, internal pacification can be understood as a "process fabricating a 'peace and security' within the social order to match the 'peace and security' imposed on colonial subjects" while "ordering the social relations of power around a particular regime of accumulation." Security is pacification.[50] As a critical concept, pacification therefore also forces us to ask questions about who is being pacified, why this is so, and for what particular objectives, while simultaneously presupposing subjects that resist efforts at their pacification.[51]

In other words, the usefulness in thinking of the politics of security in terms of pacification is that the military and police are located on a continuum of state power, aiming to order disorder with quite similar practices and hardware, as opposed to two separate spheres with different operating logics.[52] In this light, we must be careful not to fetishize the domestic police drone by framing this development as emblematic of a radical break from traditional policing mandates—the case of police drones is interesting less because it speaks about the militarization of the police, which it certainly does, but more about the ways in which it accentuates the mutual mandates and joint rationalities of war and police. Put in a slightly different way, the police drone is but one of the newest technologies that extends or reproduces the police's own design on the pacification of its territory. Indeed, the military and police are united in their mandate to pacify their respective territories and populations.[53] For certain, surveillance and intelligence-gathering, and the continual threat of violence, structure the organizational animus of not only militaries but also domestic policing – an animus moved by the "demand for order."[54] But such "order" is not only reproduced but also actively fabricated by police power.[55] Much like the police helicopters armed with powerful high-resolution cameras, flying above city streets, sidewalks, alleyways, parks, homes and lawns, unmanned vertical policing extends the police dream of pacification through air power, or a scopic verticality.

Importantly, as a technology of pacification the drone must be understood, in its logic and design, as a technology of police and not merely military power. The police drone, on this view, is not a feature of police "militarization," but a technology already structured by police logic—and here I am referring to the broad notion of the "social police" that predates yet still structures the uniformed institution now thought of as "the police."[56] That is, air power has long been a form of police power in that the inauguration of modern air power, the 1920s and 1930s, was defined by the police concept.[57] Put another way: although most histories of air power trace its origins to military power, often speaking of WWII as the crucial historical moment, Neocleous demonstrates how air power was originally conceived by its earliest proponents in Britain and the US as an explicit police technology to be used to govern in the most general sense the colonized and other "dangerous classes." More specifically, he shows how in the 1920s many of the debates taking place in metropoles concerning colonial populations framed air power as a police technology deployed to pacify indigenous peoples and fabricate order by crushing rebellions and policing minor resistances, separating the indigenous from traditional means of production, conducting aerial surveillance including land surveys and censuses, and winning hearts and mind through moral effect. Of course, air power as police power was not only discussed but actually exercised by metropoles in the "securing" of a slew of colonized territories. Perhaps most relevant to note for our discussion of police drones in the US is the 1921 "Battle of Blair Mountain," when West Virginian coal miners were aerially bombed by the private militia of mine owners, to say nothing of the military planes that were also used to conduct reconnaissance. Most recently, the 1985 bombing from a police helicopter of the headquarters of the activist organization MOVE helps in further demonstrating that air power has in fact long been a form of police power. The drone belongs to this history.

UAVs are said to better assist police with their goal of rendering illegible geographies legible from above. "An illegible society," Scott writes, "is a hindrance to any effective intervention by the state, whether the purpose of that intervention is plunder or public welfare."[58] As Ericson and Haggerty have shown, the domestic policing of insecurity, just like the military sibling, involves the collecting of information regarding a population and territory with aims of constructing "a more accurate map of the territory and a more reasonable profile of the…people who inhabit it."[59] As an "extra patrolman in the sky"[60] that is mobile and flexible, the police UAV extends the police mandate of ordering terrestrial space by technologically mediating the territorial through the aerial—"vigilant visualities"[61] take flight within a politics of verticality.[62] The drone patrol lends itself to visually ordering that which appears disordered if observed solely from eye-level, or ground- level—the terrestrial patrol is always limited by its locality. In this sense the drone is like the police helicopter. Of police helicopters, Adey writes that "Verticality implies security from the insecurities below,"[63] and this logic certainly structures the drone stare. One police official has stated drones provide "a good opportunity to have an eye up there" and that the technology provides "a surveilling eye to help us to do the things we need to do, honestly, to keep people safe."[64] In this sense, police drones are said to provide earth-bound police officers with a superior aerial vantage point to negotiate risks, threats, and disorders through the aerial distancing of subject and object. UAVs, enthusiasts therefore claim, provide much-needed public safety interventions by producing better state knowledge. This reproduces the commonsensical refrain that the state is "the knowing subject,"[65] even though what is often taken as seeing better or more—legibility—is actually itself always a partial view, or a simplification and miniaturization that excludes other forms of knowledge.[66] As Feldman states: "The circuit formed by vision and violence is itself circumscribed by zones of blindness and inattention."[67]

It is not simply a detached aerial view of an entire city that is imagined by police, but also the ability to intervene on a local level. Much like air power in combat, the police pursuit of mastering the atmosphere converges with a desire for an "unblinking eye—an omnipresent view provided by efficient UAV cycles and sequences that seeks to observe an asymmetric yet omnipresent threat with the capacity to unpredictably surprise and disrupt."[68] No matter how high the UAV soars for the police to gain an ocular superiority, it is important to remember that since the aerial view is always tethered to the ground, it is never merely ocular. Rather, it is a "vision that is practiced and touched."[69] This touch, I suggest, is realized in the culmination of a particular form of organized suspicion, namely, the manhunt.

Chamayou has recently argued that the aerial drone is the contemporary emblem of the militarized manhunt.[70] Hence the foundational structure of the "war on terror" is not a Clauswitzian duel between states, but the asymmetrical hunt for human prey. Here we could mention the quite literal hunts for Bin Laden and other suspected insurgents, Saddam Hussein after the 2003 invasion, and of course the "targeted killings" of suspected enemies on a drone "kill list," including US citizens. Key to the chase is the process of identification leading the hunter to the location of the hunted for either capture or killing, but primarily for the latter. The hunt has been a central component of pacification and accumulation,[71] and as already stated, the drone is the quintessential emblem of this new "manhunt doctrine" of contemporary warfare. The drone is the mechanical, flying and robotic heir of the dog of war. It creates to perfection the ideal of asymmetry: to be able to kill without being able to be killed; to be able to see without being seen. To become absolutely invulnerable while the other is placed in a state of absolute vulnerability. 'Predator', 'Global Hawk', 'Reaper' – birds of prey and angels of death, drones bear their names well.[72]

The drone, then, is a technology of manhunting, and this is true whether the drone in question is solely capable of surveillance or one of the "hunter-killer" drones equipped with Hellfire missiles. The drone is oriented to both the "capture" of state-produced images and the capture of those marked as Other. Historically though, the state-sanctioned manhunt has configured the animus of domestic policing more so than it has organized military violence abroad.[73] That is, the state's deployment of the manhunt has historically belonged more to police than the military. Therefore, keeping with the argument made in the previous section that air power has long been a police power, we can say that the unmanned military hunts so clearly important to the war on terror belong not only to the logic of war, but to the logic of police.

On this note, we might find it helpful to understand drones as not only a hunting technology in the service of external pacification, but a relation of domination animating the very heart of police power. A consideration of manhunting as an actual relation between dominant and dominated is to take seriously "technologies of predation indispensable for the establishment and reproduction of relationships of domination."[74] As the "state's arm of pursuit, entrusted by it with tracking, arresting, and imprisoning," Chamayou writes, policing is a hunting institution claiming a "monopoly on legitimate tracking" and capture.[75] Policing as a human hunting institution is grounded in the historical and routine workings of the police—patrolling, investigating, tracking, capturing, and even killing. Although the practice of the police manhunt is often associated with high-profile, media-driven pursuits, most recently observed with the organized hunts for ex-LAPD cop Christopher Dorner and the two Chechen-born brother suspects in the Boston Marathon bombing, we should resist seeing the hunt as only or even primarily as a form of spectacle. That is, policing as fundamentally the practice of hunting human prey is best attested to by the routine, normalized, and hence often invisible, operation of police power. As one writer for a police magazine affirms, "Law enforcement exists to keep society safe from criminals, which means apprehending and arresting those who would do harm. Police manhunts for wanted criminals are daily occurrences throughout America and Canada. Most manhunts are routine police work and garner little public attention."[76] Because the manhunt is a practice of the powerful hunting the relatively powerless, the police hunt for human prey, like all forms of manhunting, performs a far-reaching asymmetry in terms of the resources and means of tracking,[77] and this is epitomized by vertical security technologies such as the police helicopter and now the police drone.

Police drones extend the traditional police hunt in powerful ways by augmenting the grounded patrol agent with a vertical optic of advanced tracking technology. An unmanned systems editor for Janes Defence Weekly has stated that drones "could be used for anything you currently use a police helicopter for, so to follow a car chase, or to find a suspect who is hiding or for search and rescue missions. The cameras they carry can be very sophisticated, they can lock onto a car and follow it, without having someone constantly monitoring the pictures. They can then be transmitted back to police HQ."[78] One Miami-Herald journalist, perhaps unwittingly but nevertheless tellingly, articulates police drones as manhunting technology when he writes that the local police drone has the capability of "training powerful lenses on its prey."[79] The website for Vanguard Industries offers a short video promoting their Shadowhawk to "public safety" agencies that positions the viewer to see from the aerial view as the unmanned system engages in mock scenarios of the police hunt. In one scenario, titled "Tactical Night Time Ops: Officer Directed to Suspect," the viewer observes thermal imaging technology illuminating a human body hiding in what appears to be a wooded area. Over the radio we hear the suspect referred to as a target, as the drone operator guides a terrestrial officer to the precise location of the hunted. We then see the officer, silhouette illuminated with his weapon drawn, approaching the suspect as the human prey kneels to be arrested – a hunt and then a capture. Interestingly, a Monmouth University survey found that 67% of US citizens supported the use of police drones to track down "runaway criminals" and 64% supported drones policing "illegal immigration."[80]

To further push this argument we only need to consider that the move to weaponize police drones began before the exclusively surveillance variety was common in US skies. This is not all that surprising if we recall the bombings of Blair Mountain and MOVE headquarters, and more directly, the fact that military drones developed first as surveillance technologies and only later germinated into the hunter-killer drone. For example, the police version of the Shadowhawk can be armed with a taser and a stun baton. As one journalist reports: "The most relevant weapon for chasing fugitives might be the beanbag launcher. Its ammunition, though, isn't called a beanbag; it's a 'stun baton'." A Vanguard official stated: "You have a stun baton where you can actually engage somebody at altitude with the aircraft. A stun baton would essentially disable a suspect"[81]—here the coercive violence underpinning routine policing is buttressed by the capability to not only track but to literally capture with a potential debilitating blow to the hunted suspect. In a report on military UAV applications for domestic policing, two military researchers discuss a military training exercise experimenting with a "UAV non-lethal payload" that "is directly relevant to civilian police missions." Here they discuss that with little training, an individual agent was successful in dropping "smoke canisters, steel spikes for destroying tires, and propaganda leaflets, all with incredible precision."[82] Although commonly mentioned police drone "payloads" are "less than lethal" tasers, tear gas, high-pitch sound weapons, and rubber bullets, it is not hard to imagine police drones with firearm capabilities—as the non-police version of the Shadowhawk is equipped with a 12-gauge shotgun and grenade launcher that has been deployed to hunt Somali "pirates" in the Gulf of Aden. Interestingly, this move to weaponize police drones coincides with the US military's increasing emphasis on weaponizing its own micro-UAVs—as exemplified with AeroVironment's Switchblade. In South Carolina, two agencies joined forces to create a surveillance drone that allegedly can also be weaponized. According to the Sheriff: "We do have the capability of putting a weapon on there if we needed to … We could put one on there. Hopefully we would never have to use it." In the candid language of a professional hunter of humans, the Sheriff stated, "This is an example of where jurisdictional boundaries are broken down for a criminal … Quite simply put, they can't run."[83] Manhunts always risk a certain embarrassment for the state as they raise the possibility of failure or non-capture.[84] Here we can see how police drones are imagined as one possibility of reducing this potential public humiliation.

Although drones are only just now emerging as domestic policing technology and therefore unmanned manhunts exist, as of now anyway, primarily in a police imaginary, there are already concrete examples of unmanned manhunts. In what is probably the first time a police drone actively assisted in the arrest of a suspect, in 2009 the Texas Department of Public Safety used the Wasp III to assist a SWAT team in executing a search warrant on a home that they believed had weapons and drugs inside, and eventually the pursued man was arrested.[85] In 2010 an unmanned hunt took place in Britain when a vehicle was allegedly stolen and one of the two suspects successfully outran police, who claimed to lose sight of the suspect in a thick fog. Merseyside police then deployed a small drone with body heat detection: "Using its thermal imaging equipment the device quickly located its target in bushes beside the canal through his body heat and relayed live pictures to a police van nearby. Foot patrols then went and arrested him."[86] The anti-social behavior taskforce official stated: "These arrests demonstrate the value of having something like the UAV." But the aerial hunt of domestic suspects is not monopolized by the police themselves, as attested to by the fact that in 2011 a US Predator drone assisted North Dakota police in the surveillance and arrest of cattle ranchers. While looking for several missing cows on a 3,000 acre farm, the county Sheriff was chased off the property by three men with rifles. The next day a Predator drone from the local air force base was called in, along with a SWAT team and bomb squad and additional officers from nearby departments. Flying two miles overhead, the Predator's powerful surveillance system located the ranchers and discerned that they were unarmed—the three men were then arrested in a police raid.[87] Although this specific case of using military drones domestically was challenged in court, a judge controversially ruled in favor of the state.[88] Furthermore, following media reporting of this event, state authorities admitted that not only do Predator drones frequently assist this particular police department but that Predators are used in domestic investigations by the Federal Bureau of Investigation (FBI) and Drug Enforcement Administration (DEA) (Bennett, 2011). As the LA Times reports, "Officials in charge of the fleet cite broad authority to work with police from budget requests to Congress that cite "interior law enforcement support" as part of their mission."[89] Similarly, it has been reported that military drone operators in Nevada have trained by practicing their aerial tracking techniques on civilian vehicles driving on US roadways. Upon observing this firsthand, a journalist inquired: "Wait, you guys practice tracking enemies by using civilian cars?" A training exercise only, said the Air Force officer.[90]

This movement towards the unmanned hunt is also illustrated by what police say they are going to use them for. In Maryland, a police department stated in official FAA documents that a drone would be deployed to aerially monitor "people of interest (watching open drug market transactions before initiating an arrest)" as well as "aerial observation of houses when serving warrants," the searching for marijuana fields, and search and rescue missions. Similarly, an Arkansas department has stated in FAA documents that their drone is equipped with powerful infrared and zoom cameras that can pan and tilt to "track objects of interest even when the helicopter's nose is pointed away from the object." Montgomery County police's Shadowhawk will be used to "enhance and support tactical operations," such as "SWAT and narcotics operations [that] will utilize camera and FLIR systems to provide real time area surveillance of the target during high risk operations." Alabama police purchased a drone "In response to the need for situational awareness and intelligence" that will be deployed "in response to a specific dedicated law enforcement mission in a defined area" such as "covert surveillance of drug transactions" along with "pre-operational planning and surveillance, maintaining operational security, and obtaining evidentiary video." In Ogden City, Utah, a small city with just over 82,000 residents, local authorities asked the FAA to approve the use of a "nocturnal surveillance airship" that would provide "law enforcement of high crime areas" with the hope of identifying "suspicious activity." The FAA ultimately denied this request. As these examples clearly demonstrate, the police themselves articulate the police drone as first and foremost a tracking and pursuit technology – not a technology only for an abstract aerial view, but a grounded, normalized police practice of targeting. This clearly provides the police a powerful new tool to track and capture whoever it deems suspicious, yet the drone imaginary outlined above—"criminals," "fugitives," drugs, "high crime areas," "suspicious activity"—predict that policing's unmanned manhunt is predisposed to tracking and capturing the poor and downtrodden.[91]

Unmanned hunting never exists outside of the political, economic, and cultural configurations that form subjects as objects. "Seeing more only means having more suspects," as Knechtel puts it.[92] In other words, drone systems are incapable of an impartial objectivity, but rather perform a "techno-cultural production of targets"[93] where institutional mandates, cultural logics, political rationalities, and technological limits circumscribe the very rules of delivering state surveillance and violence. Today the drone is the quintessential visual prosthetic that forges political subjects asymmetrically through the narrow optics of tracking and targeting measurements and the contextual deficiencies of political economy and cultural inscriptions. Unmanning the police manhunt is loaded with violence regardless of individual drone capability as they only exist in relation to the broader organizational animus of state power.

The founding act of police was the hunting of the poor, vagrants, beggars, and the colonized.[94] This history still weighs on the present, and the drone needs to be situated within this history, a history which is, in effect, the history of pacification. Let me finish with a recent experience to highlight this point.

While I was amongst a group of police officers one day, an officer brought up how he had recently watched on CNN a police helicopter hunt down a fleeing suspect. This quickly morphed into a brief comment on police helicopters, specifically how the LAPD air units notoriously instill fear into residents. Yet this quickly then morphed into how, as one officer stated, in the near future aerial drones would be the preferred choice for providing vertical security. Another officer, echoing media reports, expressed how micro-police drones would be able to fit in the trunk of a patrol car and deployed at the officers' whim. On this, one officer joked how he would like to someday intentionally crash his hypothetical drone into what is essentially the "ghetto" part of the city to literally wipe them "off the map." The other officers laughed at the thought.

As this example suggests, and as Chamayou argues, the hunt has long induced great pleasure in those doing the hunting while the hunted prey exists and moves through space in a constant state of anxiety, largely due to the "radical dissymmetry" in the technologies of tracking. For those living under the drone stare in warzones abroad, such as Pakistan, Afghanistan, and the Occupied Territories, fear and nightmares define their experience. Although it is too early to completely understand the specific ways police drones might also induce fear and terror into citizens of the Global North, the alarm and dread produced by police helicopters in the vertical patrolling of urban space is a useful parallel that points to the affective trepidation potentially provoked in a near future with ubiquitous unmanned policing. In many ways, even those bourgeois communities and citizens usually eclipsed from the police gaze will come under the stare of unmanned policing, to the extent that air power obliterates any useful distinctions between suspect and bystander, target and non-target. As one LA journalist wrote in 1992, "Hearing LAPD helicopters circle overhead is a nightly phenomenon over much of the Los Angeles basin, even in middle-class neighborhoods like my own...the helicopters contribute to the perception that something is very wrong with this city." He continues:

> Their circular flight patterns have a way of making people feel as if they're smack in the center of a crime drama. They get under people's skin in a way that the soaring crime statistics can't ...every time the helicopters hover and circle overhead I'm reminded of my anxieties. I was insecure before the Los Angeles riots. Now the sight and sound of helicopters above compounds the tension.[95]

If this is the case for this seemingly privileged journalist, it is certainly true that the captives of wage labor, the dispossessed, perpetually hunted poor will bear the brunt of any aerially-induced terror and fear. That is, the possibility of a totalizing aerial legibility has not and will not equalize people's experience of police terror. The aerially-induced anxiety of the police helicopter hunt is best depicted in the rapper Ice Cube's song, "Ghetto Bird," where he writes "Why, oh why must you swoop through the hood like everybody from the hood is up to no good" and "Run, run, run from the ghetto bird Run." This is not meant to jeopardize the suffering of those who are "living and dying under drones" by the hunter-killer Predators and Reapers by turning their suffering into "our own." But it is to suggest that domestic policing's unmanned manhunt is also circumscribed by not dissimilar relations of domination that are generative of its own peculiar patterns of physical and psychic insecurity.

I am arguing, then, that the police drone underlines the power relations between those that are dominated and those that do the dominating, the hunted and the hunters. Within this relation stand the everyday hunters that are the police. The rise of police drones makes more perceptible this radical asymmetry between the techniques of the hunters and the hunted, or brings this relation of domination to the forefront, in similar but in even more dramatic fashion than the SWAT team or armored vehicle. As a nascent verticality organizing state suspicion, tracking, and capture, the unmanning of the police manhunt is but the newest symbolic marker of the pacification project that the poor and oppressed have been living and dying with all along. But the very notion of pacification always presupposes populations that resist and is therefore never a completed, fulfilled project.[96]

NOTES

1   Noah Shachtman, "CIA Chief: Drones 'Only Game in Town' for Stopping Al Qaeda," *Wired. com*, May 19, 209, http://www.wired.com/dangerroom/2009/05/cia-chief-drones-only-game-in-town-for-stopping-al-qaeda/
2   Jo Becker and Scott Shane, "Secret 'Kill List' Proves a Test of Obama's Principles and Will," *The New York Times*, May 29, 2012, http://www.nytimes.com/2012/05/29/world/obamas-leadership-in-war-on-al-qaeda.html?_r=2&hp&; David Cole, "Killing Citizens in Secret," *The New York Review of Books*, NYR Blog, October 29, 2011, http://www.nybooks.com/blogs/nyrblog/2011/oct/09/killing-citizens-secret/
3   Tyler Wall and Torin Monahan, "Surveillance and Violence from Afar: The Politics of Drones and Liminal Security-Scapes," *Theoretical Criminology* 15, no. 3 (2011): 239-254.
4   Allen Feldman, "Violence and Vision: The Prosthetics and Aesthetics of Terror in Northern Ireland," *Public Culture* 10, no. 1 (1997): 25-60.
5   Kevin D. Haggerty and Richard V. Ericson, "The Surveillant Assemblage," *British Journal of Sociology* 51, no. 4 (2000): 605–622.
6   Stephen Graham, *Cities Under Siege: The New Military Urbanism* (London: Verso, 2010); Michel Foucault *'Society Must Be Defended': Lectures at the Collège De France, 1975-1976*. (New York: Picador, 2003); Alfred McCoy, *Policing America's Empire: The United States, the Philippines, and the Rise of the Surveillance State* (Madison: University of Wisconsin Press, 2009).
7   Stephen Graham, *Cities Under Siege*, xix.
8   See Medea Benjamin, *Drone Warfare: Killing by Remote Control* (OR Books, 2012); Nick Turse and Tom Englehardt, *Terminator Planet: The First History of Drone Warfare, 2001-2050* (CreateSpace Independent Publishing Platform, 2012).
9   Karl Marx (1867, 1976), *Capital: A Critique of Political Economy, Vol. 1*, trans. B Fowkes. (Harmondsworth: Penguin, 1867/1976).

10 Eli Lake, "Contractors Pitch Spy Tech to Cops," *The Daily Beast*, October 18, 2011, http://www.thedailybeast.com/articles/2011/10/18/defense-cuts-force-contractors-to-look-to-sell-spy-tech-to-cops-others.html
11 Cited in Lake, "Contractors Pitch Spy Tech to Cops."
12 Ibid.
13 S. Smithson, "Drones over US get OK by Congress," *The Washington Times*, February 7, 2012, http://www.washingtontimes.com/news/2012/feb/7/coming-to-a-sky-near-you/.
14 Quoted in Joan Lowy, "Pressure Builds for Civilian Drone Flights at Home," *Yahoo News*, February 27, 2012, https://www.yahoo.com/news/pressure-builds-civilian-drone-flights-home-150120049.html
15 Christine Clarridge, "Eye-in-sky SPD Drones Stir Privacy Concerns," *Seattle Times*, April 12, 2012, http://seattletimes.com/avantgo/2018034937.html.
16 David E. Nye, *American Technological Sublime* (Cambridge and London: The MIT Press, 1994).
17 Mike Butts, "Aerial Drone Buzzes Way to Canyon County," *Idaho Press-Tribune*, April 24, 2012, http://www.idahopress.com/news/aerial-drone-buzzes-way-to-canyon-county/article_415110e4-8dd5-11e1-b648-001a4bcf887a.html.
18 Andrew Becker, "Border Agency Seeks More Unmanned Aircraft Use in Calif.," *California Watch*, January 5, 2012, http://californiawatch.org/dailyreport/border-agency-seeks-more-unmanned-aircraft-use-calif-14326; Mark Rockwell, "Northern Border Protection Drones Now Fly Almost 1,000 Miles," *Government Security News*, January 24, 2011, http://www.gsnmagazine.com/node/22283?c=border_security.
19 Stephen Dinan, "Border Agency Overextended on Drone Program," *The Washington Times*, June 11, 2012, http://www.washingtontimes.com/news/2012/jun/11/border-agency-overextended-drone-program/.
20 Butts, "Aerial Drone Buzzes Way to Canyon County," 2012.
21 Noelle Newton, "DPS Using Drones to Fight Crime Locally," *KVUE News*, January 24, 2011, http://www.kvue.com/news/local/DPS-using-drones-to-fight-crime-locally-114518554.html.
22 Nancy Flake, "MCSO Waiting for Right Moment to Deploy Drone," *Your Conroe News*, May 3, 2012, http://www.yourhoustonnews.com/courier/news/mcso-waiting-for-right-moment-to-deploy-drone/article_12c6dc67-e005-5e76-820c-043c755690d7.html.
23 CBS, "Is The NYPD Experimenting With Drones Over The City? Evidence Points To Yes," *CBS New York*, January 23, 2012, http://newyork.cbslocal.com/2012/01/23/is-the-nypd-experimenting-with-drones-over-the-city-evidence-points-to-yes/.
24 Aaron Henry, "Capital, Security, Privacy and Pacification," *Socialist Studies* 9, no 2 (2013): 94-110.
25 Andrea Stone, "Drone Program Aims To 'Accelerate' Use Of Unmanned Aircraft By Police," *The Huffington Post*, May 22, 2012, http://www.huffingtonpost.com/2012/05/22/drones-dhs-program-unmanned-aircraft-police_n_1537074.html
26 Jefferson Morley, "Drones' New Weapon: P.R.," *Salon.com*, May 22. 2012, http://www.salon.com/2012/05/22/drones_new_weapon_p_r/.
27 Ibid.
28 Ibid.
29 Mark Neocleous, *Critique of Security* (Montreal, Kingston, Ithaca: McGill-Queen's University Press, 2008).
30 Quoted in Tom Barry, "How the Drone Warfare Industry Took Over Our Congress," *Alternet.com*, November 30, 2011, emphasis mine, http://www.alternet.org/story/153278/how_the_drone_warfare_industry_took_over_our_congress
31 Allen Feldman, "Securocratic Wars of Public Safety" *interventions* 6, no. 3 (2004): 330-250.
32 Quoted in Neocleous, *Critique of Security*, 30.
33 Michel Foucault, *Security, Territory, Population*.
34 George Rigakos, *The New Parapolice: Risk Markets and Commodified Social Control* (Toronto: University of Toronto Press, 2002).
35 Mark Neocleous, " 'A Brighter and Nice New Life': Security as Pacification," *Social & Legal Studies* 20, no. 22 (2011): 191-208.
36 Neocleous, *Critique of Security*.
37 Cindi Katz, "The State Goes Home: Local Hyper-Vigilance of Children and the Global Retreat from Social Reproduction," *Social Justice* 28, no. 3 (2001): 47-56.
38 Neocleous, *Critique of Security*.
39 Naomi Klein, *The Shock Doctrine: The Rise of Disaster Capitalism* (New York: Picador), 386.

40 Katz, "The State Goes Home"; Rigakos, *The New Parapolice*.
41 Deborah Cowen and Amy Siciliano, "Surplus Masculinities and Security," *Antipode* 43, no 5 (2011): 1516-1541; Shelley Feldman, "Surveillance and Securitization: The New Politics of Social Reproduction," in *Accumulating Insecurity: Violence and Dispossession in the Making of Everyday Life*, eds. Shelley Feldman, Charles Geisler, and Gayatri A. Menon (Athens and London: The University of Georgia Press, 2011).
42 Peter B. Kraska, *Militarizing the American Criminal Justice System* (Boston, MA: Northeastern University Press, 2001).
43 Kevin D. Haggerty and Richard V. Ericson, "The Military Technocultures of Policing," *Militarizing the American Criminal Justice System*, eds. Peter B. Kraska (Boston, MA: Northeastern University Press, 2001), 43-64.
44 Guillermina Seri, *Seguridad: Crime, Police Power and Democracy in Argentina* (New York and London: Continuum, 2012), 119.
45 Simon Hallsworth and John Lea, "Reconstructing Leviathan: Emerging Contours of the Security State," *Theoretical Criminology* 15 (2011): 141-157.
46 Neocleous, "A Brighter and Nice New Life"; Neocleous, "The Police Dream of Pacification."
47 Neocleous, "A Brighter and Nice New Life."
48 See Jeremy Kuzmarov, *Modernizing Repression: Police Training and Nation-Building in the American Century* (Amherst and Boston: University of Massachusetts Press, 2012); McCoy, *Policing America's Empire*, 2009.
49 Steve Herbert, *Policing Space: Territoriality and the Los Angeles Police Department* (Minneapolis and London: University of Minnesota Press, 1997).
50 Neocleous, "A Brighter and Nice New Life."
51 George Rigakos, "'To Extend the Scope of Productive Labor': Pacification as Police Project," *Anti-Security*, eds. Mark Neocleous and George Rigakos (Ottawa: Red Quill Press, 2011).
52 Rigakos, "'To Extend the Scope of Productive Labor'."
53 Herbert, *Policing Space*.
54 Allan Silver, "The Demand for Order: A Review of Some Themes in the History of Urban Crime, Police, and Riot," *The Police: Six Sociological Essays, eds*. eds. David J. Bordua (New York: John Wiley and Sons, 1967), 1-24.
55 Mark Neocleous, *The Fabrication of Social Order: A Critical Theory of Police Power* (London and Sterling, Virginia: Pluto Press, 2000).
56 Neocleous, *The Fabrication of Social Order*; Foucault, *Security, Territory, Population*.
57 Mark Neocleous, "Air Power as Police Power," *Environment and Planning D: Society and Space* 31, no. 4 (2013): 578-593.
58 James C. Scott, *Seeing Like a State* (New Haven and London: Yale University Press, 1998), 78.
59 Richard V. Ericson and Kevin D. Haggerty, *Policing the Risk Society* (Toronto and Buffalo: University of Toronto Press, 1997), 135.
60 Jessica Miller, "Ogden Eyes Eye-in-the-sky," *Standard-Examiner*, January 13, 2011, http://www.standard.net/topics/city-government/2011/01/12/ogden-eyes-eye-sky.
61 Louise Amoore, "Vigilant Visualities: The Watchful Politics of the War on Terror," *Security Dialogue*, 38, no. 2 (2007): 215-232.
62 Eyal Weizman, *Hollow Land: Israel's Architecture of Occupation* (London and New York: Verso, 2007).
63 Peter Adey, "Vertical Security in the Megacity: Legibility, Mobility and Aerial Politics," *Theory, Culture & Society*, 27, no. 6 (2010): 58.
64 See Benjamin, *Drone Warfare*.
65 Mark Neocleous, *Imagining the State* (Maidenhead and Philadelphia: Open University Press, 2012).
66 Scott, *Seeing Like a State*.
67 Feldman, "Violence and Vision," 29.
68 Peter Adey, Mark Whitehead, and Alison J. Williams, "Introduction: Air-target: Distance, Reach and the Politics of Verticality," *Theory, Culture & Society*, 28, no. 7-8 (2011): 179.
69 Adey, "Vertical Security in the Megacity," 109.
70 Grégoire Chamayou, "The Manhunt Doctrine," trans. Shane Lillis, *Radical Philosophy* 69 (2011): 2-6.
71 Neocleous, "The Police Dream of Pacification."
72 Chamayou, "The Manhunt Doctrine."

73 Grégoire Chamayou, *Manhunts: A Philosophical History*, trans. Steven Rendall (Princeton and Oxford: Princeton University Press, 2012); Neocleous, "The Police Dream of Pacification."
74 Chamayou, *Manhunts*, 1.
75 Ibid., 89.
76 Robert O'Brien, "Police Manhunts," *Police: The Law Enforcement Magazine*, March 18, 2009, http://www.policemag.com/blog/swat/story/2009/03/police-manhunts.aspx.
77 Chamayou, *Manhunts*.
78 Jessica Elgot, "Police Drones: Unmanned Air Vehicles Could Monitor Protests, Riots And Traffic In UK," *Huffington Post*, January 10, 2012, http://www.huffingtonpost.co.uk/2012/10/01/police-drones-unmanned-air-protests-uk-riot-traffic_n_1928339.html.
79 Charles Rabin, New Police Drones Keep on Top of the Case," *The Miami Herald*, February 2, 2011, A1.
80 Monmouth University, "US Supports Some Domestic Drone Use," Monmouth University, 2012, http://www.monmouth.edu/assets/0/84/159/2147483694/3b904214-b247-4c28-a5a7-cf3ee1f0261c.pdf.
81 Michael Moore and Michael Scott, "America Edges to Brink of Armed Police Drones," *Pacific Standard*, November 16, 2012, http://www.psmag.com/legal-affairs/america-edges-to-brink-of-armed-police-drones-37837/.
82 Douglas Murphy and James Cycon, "Applications for Mini VTOL UAV for Law Enforcement," http://www.uavm.com/images/Law_UAV_for_Law_Enforcement.pdf.
83 Lauren Talarico, "A.I.R. (Ariel Intelligence and Response) to Help Law Enforcement," *Wltx.com*, March 22, 2011, http://www.wltx.com/news/article/129337/2/From-Toy-to-Life-Saving-Tool.
84 Chamayou, *Manhunts*.
85 Noelle Newton, "DPS Using Drones to Fight Crime Locally," *KVUE News*, January 24, 2011, http://www.kvue.com/news/local/DPS-using-drones-to-fight-crime-locally-114518554.html.
86 Liz Hull, "Drone Makes First UK 'Arrest' as Police Catch Car Thief Hiding Under Bushes," *The Daily Mail*, February 10, 2010, http://www.dailymail.co.uk/news/article-1250177/Police-make-arrest-using-unmanned-drone.html.
87 Brian Bennett, "Police Employ Predator Drone Spy Planes on Home Front," *Los Angeles Times*, December 10, 2011, http://articles.latimes.com/2011/dec/10/nation/la-na-drone-arrest-20111211.
88 Jason Koebler, "Court Upholds Domestic Drone Use in Arrest of American Citizen," *US News*, August 12, 2012, http://www.usnews.com/news/articles/2012/08/02/court-upholds-domestic-drone-use-in-arrest-of-american-citizen.
89 Bennett, "Police Employ Predator Drone Spy Planes on Home Front," 2011.
90 Mark Mazzetti, "The Drone Zone," *The New York Times*, July 6, 2012, http://www.nytimes.com/2012/07/08/magazine/the-drone-zone.html?pagewanted=all&_r=2&
91 Political activists have been concerned about the police use of drones to pacify protests and public dissent. And since the manhunt is always vulnerable to a reversal of roles between hunted and hunters, it is unsurprising that activists have started deploying their own counter-surveillance drones during political protests.
92 John Knechtel, "Introduction," in *Alphabet City Magazine: Suspect*, ed. John Knechtel (Cambridge: MIT Press, 2006), 21.
93 Derek Gregory, "From a View to a Kill: Drones and Late Modern War," *Theory, Culture, & Society* 28, no. 7-8 (2011): 188-215.
94 Chamayou, *Manhunts*.
95 Frank H. Strausser, "Gripes: LAPD Choppers: 'Compounding the Tension'," *Los Angeles Times*, February 23, 1992, http://articles.latimes.com/1992-11-23/local/me-554_1_los-angeles.
96 Rigakos, " 'To Extend the Scope of Productive Labor'."

The War on Purse-Snatchers
Law and Order as Pacification[1]

Deniz Özçetin

In the summer of 2013, Turkey witnessed one of the most important events in the country's history of popular protest. Anti-government protests started in Taksim Gezi Park in Istanbul and soon spread to nearly every city. If the strength, determination, and commitment of the protesters was one reason for the worldwide attention, the other was the police violence in suppressing the demonstrations. During the protests and conflicts, five people died and more than 8,000 were injured. Among the injured ones, eleven lost an eye and 104 suffered severe head trauma. In the aftermath of the events, many criminal prosecutions opened against the police. Some of those prosecutions still continue; however, in most of the cases verdicts of non-prosecution were reached. The verdicts were either justified by lack of evidence or on the grounds that police in fact did nothing wrong. The reason they are said to have done nothing wrong is because they were understood to have been acting within their discretionary power.[2]

Even though police treatment of petty street criminals does not attract as much attention as the treatment of political protesters, there are many examples of similar violent behavior. In all cases, the police have a free hand to move in a vast and ambiguous universe of discretionary powers. Since the turn of the century, these powers have been extended through many new legal regulations. The crucial thing about the expansion of the discretionary powers of the police is that it intersects and in some cases completes concomitant new legal regulations on the definition and sentencing of street crimes, especially the crimes against property. These low-level offences include pick-pocketing, purse-snatching and burglary, which are mostly associated with the urban poor. The new legal regulations provided a far more classified and detailed description of such crimes and increased their sentences. Moreover, some of them would by association come to be included under the category of 'terror crimes.' One of these crimes in particular turned out to be treated as the new symbol of urban insecurity: purse-snatching. The police authorities frequently mentioned through various media channels the purse-snatching incidents as an indicator of the allegedly increasing crime rates in big cities. Such a warning coming from law enforcement served as the ground for a call from various circles including politicians, judiciary, media, and even some academics to toughen sentences and empower the police.

Purse-snatching incidents are also mentioned along with certain neighborhoods in metropolitan areas, which are claimed to harbor crime gangs. In the media, these neighborhoods are defined as "crime nests," in which "potential purse-snatchers live."[3] The neighborhoods in question are mostly decaying inner city areas, such as Tarlabaşı and Hacıhüsrev in Istanbul, having become home to poor Kurdish migrants who make up the greatest part of the urban underclass. The neighborhoods' central location and proximity to main axes made them very valuable and subjects of urban transformation, which became an organized state policy in the 2000s. Starting from the mid-2000s, these areas became the target of successive police operations, or 'dawn operations,' conducted by heavily armed Special Forces and Riot Police officers. These operations included battering rams, sniffer dogs, and helicopters; and they were even sometimes broadcasted live on TV. The use of paramilitary tactical police units is 'normalized' by displaying the operations as if they are carried out against 'terrorist cells' and imperative for 'national security.' There are two important points to consider here: First, by including it under the category of terror crimes, the war on purse-snatching is declared as a matter of national security and the crime of purse-snatching is overtly or covertly related to the Kurdish political movement by the law enforcement mechanisms. And second, in seeking to smooth the process of economic restructuring that made urban space an actor of capital accumulation, the process can be understood as a form of pacification. Thus I argue that the recent changes in the legal regulations on the discretionary powers of the police and the penal policy can be assessed as a "pacification security job."[4]

Referring to Foucault, Neocleous argues that security should be seen as a police mechanism because it is a "constitutive technique ... mobilized in the exercise of power,"[5] and "the police project" is the fabrication of social order of capitalist accumulation through wage labour.[6] To disrupt the "colonizing and radicalizing" effects of security on discourse,[7] we should grasp security as 'pacification.' Pacification provides a more critical and fertile ground than security because it does not carry positive connotations "inextricably bound up" with the concept of security, and it registers the 'productive dimension' of the exercise of state power.[8] Thus, as the police project, pacification aims for the control and disciplining of a labour force as well as securing capital accumulation. In doing so, it includes both destructive and productive methods – to fabricate a social order "in place of existing forms of social organization."[9] The very process of capital accumulation is a *violent* process, in the sense that it includes separation of the working class from means of production and the "appropriation of unpaid labour."[10] The violence in question here constitutes the basis of 'class war'—the war of capital against labour. As Neocleous states, the class war includes both brute force and legislation, and "two of the main weapons used by the ruling class in this war are law and police."[11]

This chapter will argue that the transition to a harsher penal regime on street crimes in Turkey can be read as a pacification strategy. The new legal regulations increasing the discretionary powers of the police, expansion of both formal and informal surveillance mechanisms and tougher punishments for 'preventing future crime' are legitimized by a discourse which identifies 'war on purse-snatching' with a 'war for national security'. The pacification strategy aims to open the 'criminalized' neighborhoods of mostly Kurdish urban poor to business through urban transformation projects. To make this case the chapter discusses the recent changes in legal regulations regarding the punishment of street crimes against property, the expansion of the discretionary powers of the police, and the formal and informal surveillance mechanisms that have merged in the last decade. The argument will be that we need to read police discretion and the transition of the legal definition of petty street crime to terror activity as central to the logic of pacification.

LAW AND POLICE DISCRETION

In the *Critique of Violence*, Benjamin states that violence has a dual character: law-making and law-preserving. The law-making character of violence points to its potential for creating a new law. The very fact that the state monopolizes the use of violence renders it the 'maker of a new law.' That is why the state fears violence outside of its monopoly—for "being obliged to acknowledge it as law-making."[12] The second function of violence is law-preserving; in the case of the state, violence is used as a means to its ends. Notwithstanding that they are two different forms/characters of violence, they are somehow interlocked. As Benjamin argues, police is an institution of the modern state in which these two functions are combined in a "spectral mixture." The police use violence for legal ends; however, at the same time, it has the authority to "decide these ends ... within wide limits."[13] In that sense, the ends of police violence are not always identical with the law—rather, the 'law' of the police intervenes where the state is impotent for some reason to guarantee the ends it desires to attain. It can even be claimed that police 'invents' the law. In Benjamin's words, "therefore, the police intervene 'for security reasons' in countless cases where no clear legal situation exists." Thus police power is a "formless ... nowhere-tangible, all-pervasive, ghostly presence in the life of civilized states."[14]

Discretion can be located on the blurred borders between two functions of violence and defined as the moment when police 'invent the law.' Discretion has been acknowledged as a formal part of the police power since the establishment of the police force as a modern state institution. The idea of autonomy constitutes one of the basic pillars of discretion and provides the police officers with "quasi-judicial" powers.[15] According to the doctrine, "individual police officers have the legal right and duty to enforce the law as they see fit, including whether to arrest, interrogate and prosecute, regardless of the orders of their superiors."[16] Apart from the emphasis on 'impartiality' and 'independence from politics,' discretion is informally considered by law enforcement authorities as an 'easing behavior'—a way to create a 'law in action' to overcome many obstacles created by the cumbersome 'law in the books.' However, this capacity to create a new law has its roots in the "permissive structure of the law," which gives police tremendous powers to preserve the order.[17] 'Preserving order' turns out to be the key principle here because discretion has been a fundamental aspect of the state power in terms of political administration since the first discussions on the modern liberal state.[18] In that sense, discretion should be considered in the broader framework of administration rather than just a matter of juridicial power. The police officer, then, becomes something more than a law enforcer – he becomes a "street-level administrator/bureaucrat."[19]

Since the very act of discretion involves judgment, the police have the authority to define deviancy and the ways to take it under control. As Terrill and Mastrofski state, police officers are legally expected to "respond to citizens' 'actions' rather than 'traits' "; however, that is not usually how it happens in terms of the "law in action."[20] Thus, equal treatment of different social groups by the police becomes a political issue. In other words, there has always been a strong relation between discretion and discrimination. Reiner defines police discrimination as "unequal treatment that is not based on legally relevant criteria."[21] Discriminatory treatment usually depends on class, race and gender. Neocleous gives various examples from crime acts in the UK in the nineteenth century to underline the alliance of the discretionary values of the police with that of the ruling class. Most importantly, members of the working class have been more likely to be approached with suspicion for acting "disorderly" or engaging in "suspicious behavior."[22] Countless other studies have shown that certain characteristics such as being young, male, black, minority, and working-class increase the probability of 'being stopped and frisked, arrested, and charged by the police' as well as being treated more harshly than others during the process. Reiner refers to many studies showing that it is not possible to identify a 'factor of pure discrimination' in terms of race/ethnicity.[23] At the same time, when these groups are in the position of the "victim," it is more likely that they are "not taken seriously" and seen as "discreditable."[24]

For Alpert, MacDonald and Dunham, the earliest stage of the police discretion is the officer's "formation of suspicion."[25] Relying on field research on police-citizen encounters as well as their own observational study in Georgia, USA, they claim that minority status plays a key a role in the officers' formation of suspicion. On the other hand, the 'administrative decision making model' under which the officer acts has a strong influence on his/her behavior. This model includes internal and spatial structure of the police mechanism as well as the training and educational processes of the officers. For example, deployment of personnel can depend on the prejudices about certain areas with particular racial and ethnic compositions. Some of the neighborhoods mostly populated by migrants, ethnic minorities, and urban poor are known to have high crime rates; however, according to Alpert, MacDonald, and Dunham, they may be in a vicious circle.[26] The higher the number of police officers and the amount of surveillance in an area, the higher the arrest rate, which contributes to the discursive criminalization of the areas in question. Territorial stigmatization, in the sense Wacquant uses the term, also contributes to the harsher treatment or over-policing of the residents in those areas.[27]

Discriminatory practices are usually grounded on the extensive discretionary power of the police to stop, search and arrest in the case of probable cause or reasonable suspicion. In the US, probable cause is necessary to make an arrest; however, its scope is vaguely defined. In the legal documents, probable cause is based on the officers' knowledge of 'facts and circumstances' and 'reasonably trustworthy information.' Not surprisingly, such concepts are interpreted in various ways by the police officers and the judges.[28] Similarly, 'reasonable suspicion' grants to the police wide powers, but one rarely finds a precise description of what is reasonable. Rather, Neocleous argues that unreasonableness is defined as "a decision which is so outrageous in its defiance of logic or of accepted moral standards that no sensible person who had applied his mind ... could have arrived at it." Thus, reasonable arrests can be described as "merely one more euphemism" providing the police extensive discretionary powers among others such as "helping police with their enquiries," "obstruction," and "resisting arrest."[29]

Because police discretion results in legal disputes and lawsuits against police officers, it may seem at first glance from a liberal point of view that discretion constitutes the Achilles heel of the law enforcement structure resulting in some officers exploiting their authority. Such a perspective would tend to define discriminatory police practices as only cases of some 'rotten apples' based on psychological and behavioral explanations. However, on the contrary, discretion increases the "operational strength" of the system because it allows the police to operate in the blurred area between "law in the books" and "law in action."[30] Moreover, according to the principle of proactive policing, police can and should intervene before a crime is committed, a principle that means that they have the legal authority to decide the 'criminal potential' of certain situations. And apart from having quasi-judicial powers, the police also have the "tacit consent of the judiciary," which generally absolves them from accountability.[31]

FROM PURSE-SNATCHING TO TERRORISM

In Turkey, the debate about the definition and punishment of street crimes since 2000 has concentrated on purse-snatching. If pacification is a technique to "secure the insecurities of the bourgeois order," then the particular "insecurities" of Turkish society in the 2000s need to be addressed.[32] The ongoing war with the Kurdish groups in the Eastern and Southeastern regions resulted in mass migrations from the area to the metropolitan cities of the West in the last decades. As a result, levels of unemployment rose and jobless Kurdish migrants became more visible to the older inhabitants of the city. They settled down in the decaying urban core neighborhoods mostly identified with illegal activities such as drug trafficking, prostitution and other street crimes and stigmatized as potentially dangerous. Kurdish migrants have already been labeled as 'dangerous outsiders' or 'potential terrorists' in the public opinion and media due to nationalist sentiments, and they became fused with a 'class threat.' On the other hand, if they were employed, Kurdish migrants began to dominate an important part of the informal job market, providing the cheapest labor force.[33]

With this background in mind, we can assess the rise of 'purse-snatching' as an issue in both transition to a harsher penal policy in general and expansion of the discretionary powers of the police in particular. Until the new Turkish Criminal Code was passed in 2004, purse-snatching was not specifically defined by the law. Purse-snatchers were tried by the crime of 'stealing by distraction,' which is the simplest form of larceny. Even though in some cases courts treated purse-snatching as 'plunder' and sentenced it accordingly, the crime was generally sentenced with a couple of months or sentences were converted into fines. If purse-snatchers were children under the age of 12, then no legal procedure was involved and they were released immediately. If they were between 12 and 15, their sentence was halved.

It is a fact that some of the purse-snatching incidents ended with serious injuries or death. However, since the crime was at that point not defined specifically by the law, the deaths were treated as 'involuntary manslaughter,' and therefore the sentence was reduced to one fourth of murder sentence. In 2001, the Supreme Court decided that purse-snatchers using cars and motorcycles during the offence should be tried and punished within the frame of activities of an organized group. From that time onwards, there have been cases in which the courts treated purse-snatching as organized crime or plunder and gave high sentences. Beginning with 2002, several legislative proposals were brought to the Turkish Grand National Assembly to change the definition of purse-snatching and increase the punishment on the grounds that the crime has been on the rise recently and existing punishments were not deterrent. For example, in 2002 the MHP-Nationalist Action Party brought such a proposal. This was followed by similar proposals from the CHP-Republican People's Party and the AKP-Justice and Development Party in 2003. Hence there was a consensus in the parliament among political parties with different ideological backgrounds about the need to clarify and redefine the offence and its punishment. Such discussions were not limited to parliament but also took place in bureaucratic, judicial, and academic circles. The common point was that there was indeed a fundamental 'insecurity' in society that necessitated urgent action.

Especially after the death of a purse-snatching victim in 2004, which became the symbol of the war on purse-snatchers later, the process that ended up in legally defining purse-snatching as terrorism accelerated. In the same week of the incident, a commission of four ministers was established to investigate increasing purse-snatching incidents alongside with the problem of street children. A couple of months later, a Security Summit was held on purse-snatching to take urgent measures and "mobilize all the security forces of the country."[34] The summit was led by Recep Tayyip Erdoğan, Prime Minister of the time, and included National Intelligence Organization (MIT), police force, and gendarme. In April 2005, the Presidency of General Staff, the highest military authority of the country, prepared a 'Politics of National Security Document,' in which increasing crime rates and specifically purse-snatching were defined as an "internal threat" and argued that if necessary measures were not taken, "social problems and aggressive behavior in the society might increase."[35] It is particularly important that the document handled "public order problems" within the context of "internal migration and problems it created,"[36] linking it to the Kurdish issue and criminalizing the Kurdish migrants.

The major legal regulation on the punishment of purse-snatching that was born out of this process is the new Turkish Criminal Code that was passed in 2004 and went into effect on June 1, 2005. In addition, many other legal changes were made that amounted to a major Turkish Criminal Code Reform, including a new Law on Criminal Procedure, amendments to the Law on the Execution of Sentences and Security Measures, amendments to the Law on Misdemeanors, and amendments to the Law on the Protection of Children. Now, because the new Criminal Code was designed mainly in accordance with European Union norms and rules, many people thought it could form the basis for curtailing police authority and expanding the rights of suspects and defendants. However, despite being apparently democratizing in many aspects,[37] the new Code also redefined many offences to be included under the category of "heavy crime" and changed their penalties to harsher ones. In the case of purse-snatching, this new crime entered the law for the first time as a particular type of crime to be counted as "qualified larceny" and defined as "taking away the property carried on by special skill" (Art. 142). According to the law, the offence became punishable by three to seven years in prison, to be increased up to one third if the victim was a child, old or disabled. In the cases in which the victim is injured for refusing to give his/her belongings, the law treats the offence as plunder and gives a prison sentence from six to ten years (Art. 148). The prison sentence can increase up to 30 years depending on the degree of the injuries and can even lead to heavy life imprisonment if the offence ends up with the death of the victim (Art. 82).

Article 145 of the Criminal Code states that "Punishment to be imposed against the offense of larceny may be reduced or totally lifted if the value of the property stolen is determined to be less." This Article on the abatement of punishments is based on the value of the property and may result in the release of the defendant. However, in December 2005, Penal Department no. 6 of the Supreme Court, which is the court of appeal in plunder and purse-snatching cases, specified abatement criteria making release more difficult (Decree 2005/11360). The President of the Penal Department no. 6 argued that courts have been "misinterpreting" the Criminal Code and that led to the release of many plunder and purse-snatching defendants.[38] Another topic of discussion concerned children involved in crimes like purse-snatching. A couple of months after the new Criminal Code went into effect, regulations on child offenders were amended, and the prison sentences to children between 12 and 15 were increased (Law No. 5377, Art. 5). The Parliamentary Commission on Justice made re-arrangements on Law of Criminal Procedure in 2006 and purse-snatching was included in offences that require arrest (Law No. 5560, Art. 17). The police authorities argued that this decision would decrease purse-snatching incidents since many defendants keep doing purse-snatching during the prosecution process. The consolidation of the crime of purse-snatching occurred in 2006, with amendments to the Anti-Terror Law. These amendments treated purse-snatching crime as "terror offence" if the crime was committed within the frame of activities of an organized group (Law No. 5532, Art. 4). With that final change in 2006, it might be said that the pacification strategy was now complete: from purse-snatching to terrorism.

What has taken place in Turkey is not out of line with what has happened elsewhere: legal regulations are brought in to deal with petty insecurities of the bourgeois order, and these new laws are most obviously associated with a purported need to police the urban working class in general (and, in Turkey, the Kurdish working class in particular), said to be responsible for the minor insecurities on the street. But these minor insecurities are then easily bound up with the geopolitical insecurity that goes under the name of 'terrorism.' The Turkish Criminal Code thus plots a familiar line: from disorderliness to criminality to terrorism.[39] The purse-snatcher is not just a minor insecurity on the street but threatens the whole bourgeois order. With this in mind, we can return to the question of police discretion.

EXPANDING DISCRETION

Parallel to the changes in the penal policy, the authorities and discretionary powers of the police were extended through a series of changes: Amendments to the Law on Police Duties and Entitlements,[40] the new Law on Criminal Procedure in 2004, and amendments to the Anti-Terror Law in 2006. In the amendments to the Law on Police Duties and Entitlements in 2002, for example, it is stated that police can apprehend and use force on persons who act against his/her measures taken under the laws, resist, or prevent his/her duties (Law No. 4771, Art. 13). What counts as resistance is left to police discretion. The degree of force to be used is based on 'the character and degree of resistance' and its amount is to be increased gradually from physical force, material force (referring to handcuffs, batons, pressurized water, tear gas, physical barriers, police horses, dogs, etc.) and finally firearms to "neutralize the resisting persons."[41] The police are expected to use their discretion in deciding the degree of force to use. The same Article also states that the police must make a warning before using force. However, in the following sentence, it is suggested that in "considering the character and degree of resistance, the police can use force without making a warning." The crucial point here is that the moment when the police are going to use firearms is left completely to their discretion. In other words, the legal regulations on the use of lethal force by law enforcement do not provide any specific criteria other than the subjective dispositions of officers.

In an additional Article within the amendments made in 2007, it is stated that "There has to be a reasonable cause for the police officer based on his/her experience and impressions he/she gets from the prevalent circumstances to use his/her authority to stop a person" (Law No. 5681, 02.06.2007, Art. 1/Additional Article 4/A to Law No. 2559). Similar to the idea of 'reasonable suspicion' noted above, the law provides no clarity about this so-called 'reasonable cause' or what aspects it should possess to be 'reasonable' apart from personal experiences and impressions of individual officers. The 6th paragraph of the same Article says, "Police can take the necessary measures to prevent harm to himself/herself or others if there is sufficient suspicion that there is gun or any other dangerous material on the person or the vehicle stopped."

In 2004, a new Code of Criminal Procedure replaced the former one designed in 1929. The new law was criticized widely and heavily among many circles including the police force, politicians and the media for curtailing the authority of the police. Consequently, some articles of the code were amended in 2005. For example, in the Article on "search and seizure" the concept of "reasonable suspicion" comes back into play (Law No. 5271, Art. 116).[42] The Article holds that the police are entitled to search the body and dwelling of a person if they have a reasonable suspicion about the possibility of arrest or evidence; of course, there is no mention of the criteria of reasonable suspicion apart from discretion. It is possible to find a definition for reasonable suspicion in another legal document enacted the same year, the Regulations on the Judicial and Proactive Searches. In Article 6, reasonable suspicion is defined as "a suspicion generally felt in the face of concrete events within the course of life."[43] All told, the new legal regulations expanded the discretion of the police so widely and vaguely that the officers are nearly empowered as the sole decision-making authority to decide on stop-and-search, thanks to the concept of reasonable suspicion. In this context, amendments to the Anti-Terror Law in 2006 are once again significant. This Law also increased the discretionary power of the police, while simultaneously limiting individual rights and freedoms. As well as including crimes against property such as purse-snatching under the category of "terror crimes," with the amendment to Additional Article 2, "police officers are entitled to use firearms in the operations against terrorist organizations directly against the target without hesitation in order to neutralize the danger, in the cases of noncompliance to call for surrender or attempt to use firearms" (Law No. 5532, Art. 16).

Within the framework of these extended police powers, a new proactive policing strategy has been put into place. Proactive policing relies on the criminalization of certain social groups and their increased surveillance and control. It includes further increasing the technological capacities of police forces, centralization of intelligence networks and organization, and bringing forth an 'informing mechanism,' which puts responsibility on every citizen to prevent crime by monitoring their neighbors, colleagues, and friends, and reporting them to the police if necessary. One important feature of this proactive policing is the preparation of crime maps and surveillance strategies for the city. In 2005, an electronic surveillance system called the MOBESE (Mobile Electronic System Integration) was installed in Diyarbakır and İstanbul, and it is significant that the installation of this system was mainly justified by claiming that pick-pocketing and purse-snatching had increased. It was frequently underlined by the authorities that MOBESE would have a 'deterrence effect' on such criminals. Muammer Güler, Governor of Istanbul at the time, stated that "a criminal aware of the fact that he is being watched 24-hours will think twice before committing a crime."[44] The scope of MOBESE has since been expanded to cover local trains and buses.

Increasing police patrol on the streets was another strategy for fighting purse-snatching. Starting from the early 2000s, specially trained police forces were established to fight purse-snatching, and this effort was defined as "zero tolerance policy."[45] In 2005, the Turkish National Police announced that a Department for Preventing Street Crimes would be established in the context of increasing offences against property such as purse-snatching. In the same year, the police's 'solution pack' for the problem of purse-snatching stated that Riot Police and Special Forces Units, which until then had been "held at disposal for any possible social unrest" would now be assigned to undercover duty in shopping malls and streets to maintain security.[46] Later, they attended the police raids to some poor neighborhoods from 2006 onwards. If we bear in mind that Riot Police and Special Forces units were first found to intervene in political meetings, demonstrations, and terrorist organizations, we find once again that street crimes or neighborhood operations are treated as terror issues.

In 2007, 'Trust Teams' and 'Lightning Squads' were established in Istanbul specifically to prevent and deal with street crimes. The officers in Trust

Teams are undercover and the ones in Lightning Squads are uniformed. Trust Teams included police officers specially trained in close combat techniques and who worked undercover as shoe shiners, bagel sellers, or even drunks, and would intervene immediately by using physical force if necessary in the case of a crime. A police chief stated that the aim of the undercover Trust Teams is deterrence by "creating a 'Big Brother effect'."[47] Trust Teams and Lightning Squads were then developed further by the new Police Chief Hüseyin Çapkın and located especially in places 'where ex-convicts live' and parks and public places where 'drug-addicts' can be found. In 2011, Trust Teams and Proactive Services were turned into separate branch offices that worked under the Public Order Branch Office before to "effectively fight with street crimes." Çapkın introduced other novelties to the police force in Istanbul, defined as the "Izmir model," referring to his practices during his Police Chiefdom in Izmir. A "performance scoring system" was established within the Istanbul Police Force, based on a "carrot-and-stick" model for the policemen, in which the "successful" policemen would be awarded while the "inefficient" ones would be relocated to guard duties.[48] According to this system, crimes like theft, plunder, and purse-snatching have points for each.[49] The police officers score points per suspects they apprehend and get extra points if the suspects are arrested. What is striking in this scoring system is that the police officers would work in the 'hot spots' on their own will, without being dependent on the information given by the center on criminal incidents.[50] In other words, to increase their score, the policemen would go to 'hunt' purse-snatchers and thieves in 'suspicious neighborhoods' known for their 'suspicious residents.' There are two important things to note here. First, the scoring system that encourages the police officers to choose the area to work in at their own will with 'a propensity to illegal activities' clearly demonstrates the 'quasi-judicial powers' they possess on the 'street-level.' The police are not mere agents of law enforcement but "a form of street-level administration";[51] they 'decide' on the specific situations, specific places, and specific people to enforce the law upon. In this case, a police officer is given the full authority to patrol certain areas for any possible 'catch' and thus stigmatize the residents of the areas in question as 'usual suspects' who have the potential to disturb the social order. The second point is about the 'hunt.' As Neocleous argues, the law and the police are the "two main weapons" of the order of capital to secure accumulation through pacification.[52] While the 'law' criminalizes low-level public order offenses such vagabondage, begging, loitering, and public drunkenness, the police 'hunts' those who do or may do such activities that are obviously 'out of order'—the order of wage labour. Thus, "the police is a hunting institution,"[53] visible through various instruments including police dogs, helicopters and police drones as "unmanned manhunters."[54] In the case of the performance scoring system, the police officers are officially rendered 'bounty-hunters' hoping to be awarded with promotion.

AND THE HUNT IS WAR!

We might note, finally, the creation of a system of informers among citizens to make them a part of the police mechanism. In 2005, Turkish National Police declared that they aimed to establish a 'neighborhood watch system' within the activities of Community Policing, in which the neighborhood residents would gather on a regular basis and discuss the security issues in their area. The system would also include the neighbors monitoring each other's 'suspicious acts' and inform the police. It is argued that this mechanism would improve the informal social control mechanisms.[55] The İstanbul Governor underlines the importance of informing mechanism by saying, "Everyone should be everyone's police," and indicates that informing is a part of the "urban awareness."[56] Gathering intelligence has always been a crucial strategy of the state; the state can even be defined as an "intelligence-gathering machine,"[57] for the state's search for knowledge about its subjects is central to its power. Neocleous mentions a paranoid state of mind on the part of the state in which anyone outside of/challenging it becomes its enemy—a "security risk."[58] Thus anyone outside the social order poses a threat to its existence. Neighborhood watch systems are a part of state strategy to police the everyday life of the citizens by making them each other's police—spies of the mind of the state. In that sense, efforts of establishing an informing mechanism among ordinary citizens can be considered as part of the pacification strategy that aims to take disorderliness under control by expanding the policing mechanism beyond the borders of the state body to the 'mind and soul' of the civil society.

CONCLUSION

I am suggesting, then, that the recent changes in legal regulations in Turkey for the punishments of street crimes, expansion of police discretion, and surveillance mechanisms can be considered as the police's "extraordinary use of administrative rule-making power."[59] As Benjamin states, police organization in the modern state embodies the "spectral mixture" of law-making and law-preserving functions of violence, which can be defined as "discretion." The police have the power to 'define the deviant' thanks to their broad discretionary powers. Within that context, the so-called 'war on street crimes' provided a justification for the expansion of excessive police authority and power. The alliance of the discretionary powers of the police with that of the ruling class mentioned by Neocleous is also the case here, in terms of dislocating the urban poor from valuable locations by criminalizing them. Ambiguous statements defining the borders of police discretion allow for the stigmatization and punishment of poor Kurdish migrants as a threat to both public order and national security. For both of them, poor Kurdish migrants are at best 'suspicious' and at worst criminals, and they are turning their neighborhoods into crime nests; these areas should be cleansed and open for business.

The destruction/reconstruction dialectics underpins the pacification of Kurdish urban poor through penal policies by legitimizing the opening of their neighborhoods to market, rendering them 'safe for business.' New, rich and sterilized dwellings have been replacing poor, decayed neighborhoods harboring people working in illegal jobs or dealing with illegal business, that are definitely 'not a disciplined labour force.' As Dafnos states, "The intensification and expansion of state security apparatus is directly linked to capital's expansionist logic."[60] If we consider urban transformation projects as part of the capital's expansionist logic through increasing urban rent, we can argue that the expansion of police discretion and toughened punishments serve to secure the expansion of capital.

In this 'class war,' law is of crucial importance since it is supposed to 'restore order' to the disorderly populations and areas. According to the logic of security, just as law is supposed to bring peace internationally, it is supposed to bring 'order' domestically. However, on the contrary, war continues to rage through law since "law is pacification."[61] In other words, 'the war on street crimes' provided a rationale for the pacification of urban poor perceived as 'ungovernable,' 'out of order,' and 'in need of discipline,' and who have invaded the urban areas with a great potential for capital accumulation. Once they are absorbed by the worst paid jobs in the 'legal city,' they are no more considered as dangerous. In other words, once they became the part of the order of wage labor, the pacification process achieves one of its greatest goals. Thus, we might situate purse-snatching at the classic intersection of disorderliness and criminality, and the war on purse-snatching as "a form of low-intensity urban warfare carried out by the state in the name of security":[62] a war for pacification.

NOTES

1. I would like to express my gratitude to the other authors of this volume and the anonymous reviewers for their comments and help. I would also like to thank Mark Neocleous for supporting and encouraging me to share my work and for helping me to reassemble my ideas into a more coherent form for this paper.
2. For an elaborate list and discussion of police abuses and impunity during Gezi Park protests, see Amnesty International Turkey's reports: Amnesty International, *Gezi Park Protests: Brutal Denial of the Right to Peaceful Assembly in Turkey* (London: Amnesty International, 2013); Amnesty International, *Adding Injustice to Injury: One Year on From the Gezi Park Protests in Turkey* (London: Amnesty International, 2014).
3. "Özdemir'den polise kapkaç talimatı," *Hürriyet*, November 23, 2001, http://hurarsiv.hurriyet.com.tr/goster/haber.aspx?id=38742.
4. Mark Neocleous, "Security as Pacification," in *Anti-Security*, eds. Mark Neocleous and George S. Rigakos (Ottawa: Red Quill Books, 2011), 43.
5. Mark Neocleous, " 'A Brighter and Nicer New Life': Security as Pacification," *Social & Legal Studies* 20 (2011): 193.
6. Mark Neocleous, "The Dream of Pacification: Accumulation, Class War, and the Hunt," *Socialist Studies/Études Socialistes*, 9 (2013): 8.
7. Mark Neocleous and George Rigakos, "Anti-Security: A Declaration," in *Anti-Security*, eds. Mark Neocleous and George S. Rigakos (Ottawa: Red Quill Books, 2011), 20.

8   George S. Rigakos, " 'To Extend the Scope of Productive Labour': Pacification as a Police Project," in *Anti-Security*, eds. Mark Neocleous and George S. Rigakos (Ottawa: Red Quill Books, 2011), 62.
9   Tia Dafnos, "Pacification and Indigenous Struggles in Canada," *Socialist Studies/Études Socialistes*, 9 (2013): 59.
10 Neocleous, "The Dream of Pacification," 10.
11 Ibid., 12.
12 Walter Benjamin, "Critique of Violence," in *Reflections: Essays, Aphorisms, Autobiographical Writings*, trans. Edmund Jephcott, ed. Peter Demetz (New York: Schocken Books, 2007), 284.
13 Benjamin, "Critique of Violence," 286.
14 Ibid., 287.
15 Graham Smith, "What's Law Got to Do With it? Some Reflections on the Police in Light of Developments in New York City," in *Hard Cop, Soft Cop: Dilemmas and Debates in Contemporary Policing*, ed. Roger Hopkins Burke (Cullompton: Willan, 2004), 195-196.
16 Mark Neocleous, *The Fabrication of Social Order* (London: Pluto Press, 2000), 99.
17 Neocleous, *The Fabrication of Social Order*, 101.
18 Neocleous refers to Locke for his emphasis on the importance of discretionary power of the executive for "the good of the society" where "strict and rigid observation of the laws may do harm," Mark Neocleous, *Imagining the State* (Maidenhead: Open University, 2003), 44. Locke also implies that law may be cumbersome when dealing with 'Accidents and Necessities' that require immediate action, Mark Neocleous, "Security, Liberty and the Myth of Balance: Towards a Critique of Security Politics," *Contemporary Political Theory*, 6 (2007): 135.
19 "Street-level bureaucrats" is the name given to the public officers who "interact directly with citizens in the course of their jobs, and who have substantial discretion in the execution of their work" by Michael Lipsky. They "deliver benefits and sanctions, structure and delimit people's lives and opportunities," Michael Lipsky, *Street-Level Bureaucracy: Dilemmas of the Individual in Public Services* (New York: Russell Sage Foundation, 2010), 3-4. The police are among the "typical street-level bureaucrats." As Seri states, the police, like any other street-level bureaucrat, are able to "transform the spirit of the laws" thanks to their discretionary powers, Guillermina S. Seri, "Police Discretion as Unwritten Law: Governing the State of Exception," in *Police Discretion and Law*, ed. Radha Kalyani (Hyderabad, India: AMICUS, ICFAI University, 2009), 3.
20 William Terril and Stephen D. Mastrofski, "Situational and Officer-Based Determinants of Police Coercion," *Justice Quarterly*, 19 (2002): 217.
21 Robert Reiner, "Police Research in the United Kingdom: A Critical Review," *Crime and Justice*, 15 (1992), 478.
22 Neocleous, *The Fabrication of Social Order*, 100.
23 Reiner, "Police Research in the United Kingdom," 480.
24 One of the 'discriminated' victim categories is women. Reiner argues that the police treat women from within a "madonna/whore sexual imagery" (Reiner, "Police Research in the United Kingdom," 479). While the women considered to be in the first category are deemed worthy of a 'chivalrous' attitude, the second group usually face with a harsher treatment. In the same manner, the police choose not to intervene in most cases of domestic male violence against women. As Neocleous states, police discretion involves both over-enforcement and under-enforcement of the law (Neocleous, *The Fabrication of Social Order*, 100).
25 Geoffrey P. Albert, John M. MacDonald, and Roger G. Dunham, "Police Suspicion and Discretionary Decision Making during Citizen Stops," *Criminology*, 43 (2005).
26 Alpert, Macdonald, and Dunham, "Police Suspicion," 408, 411.
27 Loïc Wacquant, "Territorial Stigmatization in the Age of Advanced Marginality," *Thesis Eleven*, 91 (2007).
28 Seri, "Police Discretion as Unwritten Law."
29 Neocleous, *The Fabrication of Social Order*, 103.
30 Ibid.
31 Seri, "Police Discretion as Unwritten Law," 4.
32 Neocleous, "A Brighter and Nicer New Life," 92.

33   Similar to what Saborio defines as the 'paradoxical stereotype' about the urban poor living in the favelas, the Kurdish migrants are considered dangerous living in their neighborhoods but not dangerous in the worst jobs in the 'legal city,' Sebastian Saborio, "The Pacification of the Favelas: Mega Events, Global Competitiveness, and the Neutralization of Marginality," *Socialist Studies/Études Socialistes*, 9 (2013): 133.
34   "Kapkaç terörüne neşter zirvesi," *Hürriyet*, February 26, 2005, http://hurarsiv.hurriyet.com.tr/goster/haber.aspx?id=299655.
35   "İşte askerin güvenlik önerisi," *Hürriyet*, April 27, 2005, http://www.hurriyet.com.tr/index/ArsivNews.aspx?id=315041.
36   Gencer Özcan, "Milli Güvenlik Kurulu," in *Almanak Türkiye 2005: Güvenlik Sektörü ve Demokratik Gözetim*, ed. Ümit Cizre (İstanbul: TESEV, 2006), 35.
37   The major democratic reforms brought about by the new Criminal Code include regulations on stopping systematical torture, abolishment of death penalty, curtailing the authority of the police to use firearms, regulations on crimes against personal immunity and privacy, regulations on crime of thought and freedom of expression.
38   "Tahliye etmek yanlış," *Hürriyet*, December 3, 2005, http://www.hurriyet.com.tr/index/ArsivNews.aspx?id=3594854.
39   Neocleous, "A Brighter and Nicer New Life," 200.
40   The Law on Police Duties and Entitlements was enacted in 1934 and is still in effect though being subject to many changes and amendments over the years.
41   The power to use force and firearms was regulated in the Article 16 and Additional Article 6 of the law. There have been no amendments in Article 16 since 1985. In addition to the vague terms such as "breaking resistance" to define the cases in which police are entitled to use force, the Article defines conditions of entitlement to use firearms in very broad and vague expressions. The police are entitled to use force in cases of "self-defense," "where physical and material force are not adequate to break resistance," "apprehending persons with an arrest or detention warrant" and "apprehending a suspect in the act."
42   The article goes as: "In cases where there is reasonable suspicion that he may be arrested without a warrant, or evidence of the crime may be obtained, then a body search and a search of the belongings, or a search in the dwelling, business place and in the other premises of the suspect or the accused may be conducted."
43   If this expression is not enough to satisfy anyone, the article continues with giving some details which are supposed to clarify the term: "Reasonable suspicion should be based on behavior and attitude of the person related to the place to be searched. There have to be indications within the reasonable suspicion, which support the informing or complaint. The suspicion on the issues mentioned must be based on concrete facts. There has to be concrete facts which lead to the prediction of finding a certain thing or apprehending a certain person."
44   "İşte MOBESE'nin şifresi," *Hürriyet Pazar*, June 19, 2005, http://hurarsiv.hurriyet.com.tr/goster/haber.aspx?id=328290.
45   In this period, Hasan Özdemir, Istanbul Chief of Police of the time, frequently used the term "zero-tolerance" in his statements on the "war on purse-snatching." Even though there is no explicit reference to Giuliani's policies in New York, which can be defined as the first implementation of zero-tolerance policing based on proactive policing of low-level public order offences such loitering, public drunkenness, and panhandling, Özdemir most likely became the spokesperson of a new policing strategy in Istanbul Police that took Giuliani's as an example.
46   "Emniyetten kapkaça çözüm önerileri," *Hürriyet*, February 26, 2005, http://hurarsiv.hurriyet.com.tr/goster/haber.aspx?id=299654.
47   "Bu da canlı MOBESE," *Hürriyet*, June 01, 2007, http://www.hurriyet.com.tr/gundem/6625810.asp?gid=48.
48   Rewarding the policemen catching thieves, purse-snatchers, and pick-pockets is a common practice and dates far back then Çapkın's term in İstanbul; however, Çapkın systematized and standardized the rewarding mechanism and used it as a part of the appointment criteria for the police officers as he did in İzmir, Zeynep Gönen, "Giuliani in Izmir: Restructuring of the Izmir Public Order Police and Criminalization of the Urban Poor," *Critical Criminology*, 21 (2013): 93. For example, a news report from 2001 with the title "Prize Purse-snatcher Hunt" mentions that the policemen are "rewarded with 50 million liras per purse-snatcher," and thanks to that practice, purse-snatching decreased to a great extent. The news report invokes the image of bounty hunters, "Ödüllü kapkaççı avı," *Hürriyet*, March 05, 2001, http://www.hurriyet.com.tr/index/ArsivNews.aspx?id=-230112.

49 Police officers get 100 points for catching Molotov cocktails, 20 points for looting, purse-snatching, homicide, theft from house, car theft and theft from car, 15 points for attempted looting, purse-snatching, theft from house and car theft, pick-pocketing and theft from workplace, 10 points for drug-dealing, pick-pocketing from shops, fraud and fraud in money exchange, 7 points for stealing motorcycles and bicycles, all kinds of attempted theft and buying drugs, and 6 points for possession and usage of drugs, cybercrimes and unauthorized guns, "970 polise bonus tayin," *Hürriyet*, June 14, 2010, http://www.hurriyet.com.tr/gundem/15014209.asp.
50 "İstanbul Emniyeti'nde Çapkın devrimleri," *Hürriyet*, July 29, 2009, http://www.hurriyet.com.tr/gundem/12172771.asp.
51 Neocleous, *The Fabrication of Social Order*, 102.
52 Neocleous, "The Dream of Pacification," 12.
53 Grégoire Chamayou, *Manhunts: A Philosophical History*, trans. Steven Rendall (Princeton, NJ: Princeton University Press, 2012), 89, quoted in Neocleous, "The Dream of Pacification," 18.
54 See, Tyler Wall, this volume.
55 "Artık 175 bin polisin cebinde sanığa okunması zorunlu olan Haklar Bildirgesi var," *Hürriyet Pazar*, March 20, 2005, http://www.hurriyet.com.tr/index/ArsivNews.aspx?id=304962.
56 "300 okulun önüne kamera takılacak," *Hürriyet*, January 13, 2006, http://www.hurriyet.com.tr/gundem/3779581.asp. In 2007, the Social Ethics Association (TED) prepared an "Active Citizenship Project," to improve the informing mechanism within the society. It is argued that informing plays an important role in punishing the criminals, and especially crimes like purse-snatching and plunder will decrease if it is improved, including monetary rewards to the informant, "İhbarcıya para ödülü verelim," *Hürriyet*, March 21, 2007, http://www.hurriyet.com.tr/ankara/6165951.asp.
57 Neocleous, *Imagining the State*, 49.
58 Ibid., 61.
59 Neocleous, *The Fabrication of Social Order*, 102.
60 Dafnos, "Pacification and Indigenous Struggles," 65.
61 Mark Neocleous, "War as Peace, Peace as Pacification," *Radical Philosophy*, 159 (2010): 14.
62 Neocleous, "A Brighter and Nicer New Life," 203.

'The Free and Secure University'

Biriz Berksoy

University and college campuses worldwide frequently host student movements and have played important roles as "major catalysts for the initiation ... of political insurgency and social change."[1] As educational institutions, they are also spaces in which the state constitutes its presence, rationality, and power.[2] Since late 1970s, sweeping changes caused by neoliberalization processes have affected many of them through the restructuring of higher education and rearrangement of campuses. Especially during the 2000s, several educational spaces have been reorganized to include extensive surveillance mechanisms and policing assemblages in new forms that point toward a novel arrangement of networking in the subjugation of student opposition incorporating the activation of anti-terrorism regulations and the judiciary.[3]

University campuses in Turkey have undergone a similar process within the same period. The state had always been vigilant to repress student movements at universities. To reinforce its sovereignty, it had utilized physical violence as its key technology of power. However, by mid-2000s, it started to employ new power technologies prioritizing extensive intelligence gathering and infliction of psychological violence through networks constituted by state and non-state agencies. It also started to activate heavy penal courts by labeling students as members of crime/terrorist organizations. Since late 2000s, students who are active within the Kurdish liberation movement and aim to replace the relations of exploitation and domination in the Kurdish region (a substantial part of south/eastern Anatolia, a.k.a. northern Kurdistan) with an alternative political-economic web of relations have been their primary targets.

This chapter explores the recent establishment and functioning of *surveillance-policing networks* at university campuses as part of the state's changing power technologies against social opposition through an analysis of their development in Turkey during the 2000s. The main arguments of the paper are as follows: While the institutions of higher education are important apparatuses reproducing social classes,[4] they, as environments accommodating social movements, are also important sites of *pacification*. The logic of pacification is "about the shaping of the behavior of individuals, groups and classes, and thereby ordering the social relations of power around a particular regime of accumulation."[5] The growing body of work on pacification has shown[6] that core to this pacification is the process of depoliticization, which incorporates the crushing of social opposition. Examining the reconfiguration of campuses in Turkey during the 2000s—that occurred similarly to those in countries such as the US, the UK—discloses the pacification/depoliticization processes taking place within higher education institutions. Their reconfiguration has materialized within the neoliberalization process incorporating the development of a 'security state' along with its distinct governmental technologies. It arguably led to the establishment of networks among state and non-state agencies to meticulously surveil, police, and pacify the student opposition through enhanced technologies of surveillance and novel modalities of power that incorporate post-disciplinary preemptive[7] ones. By examining the case in which authorities tried to subsume Kurdish students to the existing relations of power, it is *also* possible to observe the essential role played by the law in the pacification process.

THE "SECURITY" AND DEPOLITICIZATION OF THE CAMPUS

The institutions of higher education have been subjected to critical scrutiny since 1960s and their image as institutions providing "socially neutral" knowledge was gradually repudiated.[8] Recently, an increasing number of studies on higher education problematizes the new forms of surveillance/policing mechanisms established at campuses.[9] Several of them focus on the University of California system, which constitutes a paradigmatic example especially after 2011 when the outrage of the students against privatization was aligned with the Occupy movement.[10] In these campuses, student dissidents have been subjected to different forms of violence as part of campus "security measures."[11] The use of legal prosecution against student and faculty dissenters, increasing collaboration of university administrations with the CIA and FBI, extension of surveillance to students' Internet usage, establishment of a "synopticon,"[12] blurring of the categories of "protestor" and "terrorist" via new legal regulations, e.g. the US Patriot Act,[13] and the increasing usage of police violence against students are components of these new 'security measures.'

Analyzing the techniques employed for the repression of students' revolt against the neoliberal accumulation regime and its effects on higher education, these studies disclose that campuses are sites in which mechanisms are established to dissolve resistance acts that can challenge the core elements of the social order upheld by the state. They reveal that they are "geographies of power,"[14] spaces saturated with a police logic. The spaces of higher education are, in fact, part of institutional geographies and therefore, they are "material built environments. ... which seek to restrain, control, treat, 'design' and 'produce' particular ... human minds and bodies."[15] Furthermore, as Lefebvre argues, these spaces are designed as "appropriate spaces" "for a use specified within the social division of labor and supporting political domination."[16] Accordingly, particular values, attitudes, and practices are promoted, while others are marginalized. Yet, they are also sites generating critical perspectives, counter-hegemonic projects and student grievances related to broader social demands. As empirical cases demonstrate, they are locations of contest[17] over the reconstruction of the web of social relations, and students confront strategies that aim to bring them in conformity with state policies. Hence, by furthering these studies, it can be suggested that campuses are significant sites as they are also spaces of *pacification*. They are sites in which one can observe the modalities of state power employed to disperse resistance against social order resting on a specific regime of capital accumulation and a political configuration producing "ideal citizen-subjects of capitalism."[18] Thus, one can witness at campuses the "de-politicization" methods in operation that lie at the center of the pacification process and include the crushing of social resistance against state strategies. De-politicization at campuses is executed mostly through "security measures." This reveals that "security" is, in fact, policing; it is a political technology deployed in the exercise of power.[19] Moreover, it is also possible to decipher in these struggles the *modifications* in state technologies tailored to pacify social opposition. Indeed, new technologies of the state became visible at campuses in late 2000s, which have been securitized in new forms in many countries such as the US, the UK, and Turkey. New mechanisms that can be identified as *surveillance-policing networks* have appeared at many of them.

The emergence of surveillance-policing networks within the spaces of higher education institutions in these countries is arguably one of the effects of the neoliberalization processes that has been unfolding since late 1970s. The neoliberal accumulation regime and regulatory frameworks geared to the needs of especially financial capital[20] have transformed higher education. They initiated the opening up of universities towards markets and interests of business. These institutions have turned into some form of private or quasi-private enterprises. As the public subsidies decreased, especially for arts and social sciences programs, they have increasingly relied on research contracts made with private enterprises. Thereby, while prospects for generating critical thinking are undermined, they partly or wholly turned into corporations specializing in the production of research as a commodity[21] and selling higher education through increased tuitions.[22]. As these policies have caused student resistance movements, the latest examples of which emerged in Spain, Greece, the United States, it can be argued that this process also entailed the re-configuration of these spaces with new "security measures"[23] to pacify them in line with the neoliberal restructuring of the penal spheres with fortified police organizations and booming prison systems since late 1970s.[24]

It can further be argued that the transformation of university campuses is particularly shaped by a specific state form ripening within these processes. As

Poulantzas argues, class struggles in the power block and shifts in hegemony from one class fraction to another shape the state's organizational framework and lead to the domination of one particular apparatus over others.[25] Based on this understanding, it is possible to notice the ripening of a particular state form in countries where the neoliberalization processes gained institutional depth under the ascendancy of financial capital (e.g. the US, UK, Canada, Turkey). This state form, which can be identified as a "security state,"[26] has arguably similarities with "authoritarian statism" as defined by Poulantzas. They rest on a power axis dominated by the presidential and governmental executive. The security state incorporates what Poulantzas had formulated for authoritarian statism: "Always of a more or less fictitious nature, the already greatly reduced separation of legislative, executive and judicial powers in the bourgeois state is itself subject to final elimination."[27] While within this new state form, the parliament has lost its power to the executive and the judiciary has come under the domination of the police organization.[28] Thereby, it has become the apex of the "radical decline of the institutions of political democracy and with draconian and multiform curtailment of so-called 'formal' liberties."[29]

The 'security state' has arguably a distinctive dependence on neoliberal governmental rationality. It functions through technologies for the prevention of "risks"[30] to achieve cost-reduction and efficiency.[31] As this rationality constructs every individual—including the homo *criminalis*—on the basis of homo *economicus* acting on cost/benefit calculations, it leaves an important effect on the penal sphere. It attributes "criminal potentiality" to every individual and spreads "risk" throughout the population.[32] It universalizes suspicion under which everyone is a suspect on a continuum representing different degrees of risk.[33] As Dillon argues, the universalization of risk was initiated with the globalization of finance, trade, and manufacture that transformed risk from one form of calculation into a new order of governance in the last 30 years.[34] The state of this new order attempts to prevent risks through criminalization, especially *preemptive* criminalization, when confronted with social problems such as poverty, "advanced marginality" or even social opposition.[35] It intervenes "in advance" to avoid even the possibility of an unwanted incident.[36] Hence, this state form operates through a security *dispositif*[37] steered by a logic of "preemptive security."[38] Consequently, the penal sphere has come to include several components: policing assemblages including non-state actors, reorientation of criminal prosecution away from suspect rights, establishment of massive surveillance mechanisms, and adoption of vague anti-terror legislation amenable to a widening use etc.[39] Even before 9/11, bureaucracies and technologies of prevention had been established for governing through security.[40] However, after 9/11, the adoption of new anti-terrorism acts incorporating the vague notion of "intention"[41] led to the exceptional status of Guantanamo Bay, shifting the burden of proof to the defendant,[42] the criminalization of the social opposition with terrorism[43] etc.

As these strategies facilitated the pacification of social opposition through anti-terrorism legislation, they also revealed that law has an important role in the organization of power. As Poulantzas argues, law is the code of organized public violence. It is a constitutive element of the politico-social field.[44] But furthermore, it is also a significant medium through which different modalities of power can be activated for shaping individual and collective conduct.[45] Therefore, it can be argued that the frequent usage of "anti-terrorism acts" must be evaluated not as a sign of entering into a state of exception but as the construction of a "new normalcy" in which the logic of "preemptive security" operates through the employment of new legal regulations and frequent activation of courts. Relying on this background one can decipher the transformation taking place at university campuses in Turkey.

THE SECURITY STATE IN THE UNIVERSITIES OF TURKEY

The reconfiguration of university campuses in Turkey is arguably unfolding in connection to the institutionalization phase of the neo-liberalization process in Turkey since the early 2000s. In this process, the previously existing "national security state" dominated by the militarily focused National Security Council has given way to another state form.[46] This new form can be recognized as a "security state" due to the new power axis established within the state by the Justice and Development Party (JDP) governments since 2002 and the governmental technologies put into use since 2007.

The new power axis has been founded on a series of institutional and legal modifications throughout the 2000s. During three terms of government (2002-2007; 2007-2011; 2011-2015), the JDP converted the National Security Council into an advisory board to meet the membership requirements of the European Union. It put into effect the new Turkish Criminal Code and the Criminal Procedure Code in 2005 and gained control of the Presidency and Higher Education Council after 2007. It amended key laws such as the Anti-Terrorism Law in 2006 and Police Duties and Authorities Act in 2007 and acquired the ability to re-structure the high judiciary organs through the 2010 referendum. These changes led to the formation of a new power axis under the control of the government. In effect, the police power has been enhanced as the primary channel for governing everyday life, ably supported by the judiciary.[47]

The political rationality of the JDP governments is characterized by religious conservatism, a modified nationalism recognizing ethnic differences in return for political docility, and a statist authoritarianism based on an organic vision of society.[48] Their aspiration to construct a "democratic market society" led to the extension of market rationality to social spheres. This process has included the reorganization of state institutions through neoliberal governmental technologies and reconstruction of individuals around the image of homo *economicus*. It is possible to trace the effects of this reorganization within the penal sphere. First of all, the police organization has gone through transformation in its missions and structure. In the first "strategic plan" prepared after the adoption of the Law on the Financial Management and Control of Public Institutions in 2003, its missions were articulated to a newly constructed image of *homo criminalis* who acts out of ambition/selfishness whenever s/he finds the opportunity.[49] While this image extends the potentiality for committing crime to the whole population, the core police mission is redefined as intervention in the environment to prevent crimes before they materialize. Under the impact of this logic of "preemptive security," the police power's intelligence-gathering remit has come to the fore operating through enhanced surveillance mechanisms[50] that include cooperation mechanisms established with state institutions and non-state actors. Arranging the environment with more police presence, identification of "crime zones" and drawing "threats" into the penal sphere became major goals for the police.[51]

The judicial sphere has also gone through modifications. Downgrading economic matters to technical matters, highlighting political stability, and criminalizing social struggles, the JDP governments made modifications in legal regulations to enable pulling into the penal sphere political opponents who effectively challenge state strategies and constitute "risks" for the state.[52] The most important changes are realized with the adoption of the new Turkish Criminal Code (particularly Articles 220 and 314) and the amendments made to the Anti-Terrorism Law in 2006 (Article 7). With the help of these articles as well as Article 2 of the Anti-Terrorism Law, it has become possible to prosecute a person as a member of an "armed/terrorist organization" based solely on the claim that s/he promotes its aims or supports it. These legal regulations enable the judicial treatment of political activists as members of "terrorist organizations."[53]

Now, the point to note is that the effects of this transformation can also be observed in universities and their campuses. They materialized as the extension of the neoliberalization process through which the higher education in Turkey was restructured according to market rationality. The process started in mid-1990s[54] and culminated in the initiation of a process called "Efforts for the Restructuring of Higher Education" in the late 2000s, which includes the adjustment of higher education to business needs, reconstruction of governing bodies to incorporate the business world, disciplining academic staff through performance criteria, and so on.[55]

Within this process, university campuses have also been reconfigured by means of new "security measures." They incorporated the employment of private security guards on campuses (after the enactment of Law on the Services of Private Security Guards in 2004), deployment of CCTV cameras, assignment of police officers to various tasks such as distributing leaflets and introducing "social responsibility" courses besides their main missions.[56] These strategies culminated in the preparation of the "Free and Secure University Project" by the second JDP government in 2008. The very title of the project, which brings the concepts of freedom/liberty and security together, reveals the liberal mode of thought that generated the project. As Neocleous demonstrates "behind this ... mode of thought lies a set of fundamentally illiberal justifications for a range of extreme and dangerous security measures."[57] Accordingly, the project urges the re-ordering of campuses with invasive "security measures" enabling the neutralization of any activity deemed out of "order" and disclosing the fact that "security and oppression are the two sides of the same coin."[58] It urges university rectors to demand uniformed and undercover police officers to be officially employed at campuses, to establish "guidance units" to identify "problem causing students," to appoint "security coordinators" who would participate in "security meetings" to be regularly held by the Higher Education Council, to extend the CCTV camera systems, to introduce electronic mechanisms for identifying finger prints etc.[59]

Lately arranged "security measures" reveal the state's aspiration for finding new and efficient modes in achieving the pacification of student opposition that effectively challenge the "social order." The state had always been vigilant to pacify students. Yet, surveillance mechanisms were weak. The police did not have the capacity to collect extensive intelligence. Judicial investigations and court trials were rare. The state could not penetrate the daily lives of students. It was relying on arbitrary detentions and physical torture.[60] The shift in the state form and governmental technologies have arguably brought in these new "security measures" which incorporate the judicial sphere. In fact, while the abovementioned project was being devised, hundreds of students have been taken under detention and systematic prosecutions have been initiated against them under the new Turkish Criminal Code and the amended Anti-Terrorism Law.[61] They have been prosecuted in at least 19 cases in heavy penal courts and many of them have been punished with long prison sentences. It is estimated that around 90 percent of those students pulled into the judicial sphere have been Kurdish students.[62]

PACIFYING THE KURDISH LIBERATION MOVEMENT AT CAMPUSES

At the heart of the new pacification are the Kurdish students whose movement at universities originated in 2001 when hundreds of students started petition campaigns to have Kurdish language courses in their curriculums. It was launched as part of the Kurdish liberation movement, which has as its foundation the armed struggle initiated by the PKK (*Partiya Karkerên Kurdistan*, Kurdistan Workers' Party) against the state in 1984 for the establishment of an independent Kurdistan. While the liberation movement as a strong political struggle has turned into a mass mobilization among Kurdish people, especially through the acts of civil disobedience during the 2000s,[63] Kurdish students joined it from within universities.

The liberation movement is the last one of numerous rebellions that started in the nineteenth century against the centralization efforts of the Ottoman state that developed into a "colonization process" in Kurdistan.[64] The state during the Republican period (1923-) upheld these strategies by creating an environment of structural violence in the region on political, economic, and cultural terms. The region has been subjected to uneven capitalist development and poverty.[65] The ethnic identity of the Kurdish people was not recognized until late 1990s. They were subjected to assimilation policies through physical and cultural violence. Kurdish language was banned, and extrajudicial killings, infliction of torture, and village evacuations were rampant from the beginning of the 1980s through the 1990s.[66]

The Kurdish liberation movement abandoned the goal for independence in late 1990s and endorsed an alternative standpoint for the politico-economic reorganization of the region. Through recently established organizations such as the Union of Kurdistan Communities (*Koma Civaken Kurdistan*, [*Kürdistan Topluluklar Birliği*], KCK, 2005), Democratic Society Congress (*Kongreya Civaka Demokratîk*, [*Demokratik Toplum Kongresi*], DTK, 2007) and the political party affiliated with the movement (the most recent one is Peoples' Democratic Party, *Halkların Demokratik Partisi,* [HDP]), the models of "democratic confederalism" and "democratic autonomy" have been adopted. They are based on an alternative vision of political organization without state formation, economic organization dismantling capitalist relations (including ecological exploitation), and social relations free of class and gender domination.[67] This new conceptualization of social relations, which is, in bits and pieces, put into implementation within the provinces governed by the mayors affiliated with the HDP, prioritizes the democratic self-government of people through a network of interdependent local democratic assemblies (village assemblies, local parliaments, larger congresses) functioning through direct-democracy.[68]

This vision of an alternative political-economic organization is in clear contrast to the highly limited project prepared by the JDP government in 2009 for the solution of the Kurdish question. The government project is titled "Democratic Initiative Process: The National Unity and Brotherhood Project" and mainly based on the de-politicization of the Kurdish identity,[69] the maintenance of the web of relations based on capitalist exploitation, and political domination while recognizing minor cultural rights of Kurdish people.

Kurdish students within the Kurdish liberation movement are struggling basically for the right to construct an alternative web of social relations in the Kurdish region through decentralization. Protests are organized at significant dates such as the ones on which Kurdish students were killed by the police. They hold demonstrations, put up ("unauthorized") posters, open stands, and make public statements at campuses. They became the primary target of harassment as they effectively challenge state power with their demands for an alternative political-economic order. They are, by definition, a substantial "risk" for the state, especially in a conjuncture in which negotiations are conducted with the PKK.[70] Therefore, their experiences at campuses reveal the workings of the state's new strategies for the pacification of student opposition[71] and social opposition in general.

To be able to map these strategies, interviews were held by the author between May 2013 and February 2014 with seven politically active Kurdish students (Student A, B, C, D, E, F, G) introduced by a member of the "Initiative for Solidarity with the Students in Prison" as well as with seven private security guards (PSGs) and two former students who had been politically active in a leftist student movement during the 1990s.[72] Based on these interviews, it can be concluded that at university campuses, the PSGs, university administrations[73] and the police have recently constituted close cooperation mechanisms. The police try to extend these mechanisms to students' families and their fellow students. During the interviews, almost all students indicated that the police make phone calls to their families or visit their homes.[74] They try to employ fellow students as informers. The outcome of this set of connections can be identified as *surveillance-policing networks* that operate under the domination of the police organization. They function to govern the student dissenters by suppressing alternative policy suggestions, disciplining them within the neoliberal university, and neutralizing/eliminating the recalcitrant ones as "risks."

Within these networks, the main concern is the collection of extensive information/intelligence. The PSGs collect intelligence by patrolling the campuses. Both the police and PSGs keep a record of every student activity. Extensive information is received from the camera systems. Additionally, the police record all student activities on film. Student D stressed that the police follow them even to student housing sites. They employ wiretapping and audio surveillance. The university administrations provide data to the police about the forthcoming student activities. Based on the intelligence gathered, the PSGs and the police identify the students taking part in the organization of activities as the main targets.

By utilizing the extensively accumulated intelligence/information, various strategies for the pacification of the politically active Kurdish students are implemented. Most of them materialize at the intersection of repressive, disciplinary, and post-disciplinary/preemptive modalities of power. Special emphasis is put on the infliction of psychological violence as a novelty rather than physical violence. It constitutes an important element in destroying/repressing students' wills and disciplining them, especially the politically leading ones. It is inflicted in various strategies such as staring during demonstrations, making verbal assaults, initiating disciplinary investigations, performing spectacular police raids, engaging in psychological manipulation under detention and initiating court cases. As mentioned above, some of these strategies do not only operate as a disciplinary feed-back mechanism for self-governance but also function to eliminate some students and preempt their activities because they can be "associated to a particular risk-population and because they *might* well cause a disturbance, irrespective of whether this would really be the case."[75]

As these networks overtly infiltrate into the deepest corners of everyday life at campus and are connected with the outside policing mechanisms, they transmit to the students the message that they constitute a ubiquitous and omniscient continuum. All the students recounted during interviews that they have been under the constant watch of the PSGs and the police[76] at campuses and in demonstrations and have been subjected to their verbal assaults. For example, Student B told that one day undercover police officers verbally attacked him at the school gates by saying: "How long lifetime should we cast for you? Let him wander around for two more weeks." The PSGs let the police know about the arrival of certain marked students through their explicit announcements at the gates, which can also be heard by the students. Furthermore, the police also inflict psychological violence on the families. They tell them that their son/daughter is hanging with the wrong people and if they do not do anything about it, they may not be able to see their child for a long time. They try to activate the family through stimulating emotions of guilt or threats. These strategies have an important capacity for disciplining both the politically active *and* inactive students. They urge all the students to self-regulate themselves and to stay within the boundaries set by the state.

Disciplinary investigations are also used as a factor of psychological violence and intimidation. But they are also employed for the elimination of "risks" by suspension/expulsion. A report written on disciplinary investigations conducted in Dicle and Istanbul Universities reveals that there is a steady increase in the number of investigations.[77] It also demonstrates a hierarchy between the politically active students. The students who face investigations first are Kurdish students, and then the leftist students are investigated, especially if their organizations have become more active at the time. Kurdish students are criminalized via news broadcasted in the Kurdish media to establish links between student activism and the PKK. According to the report, investigations are generally initiated with a demand made by the police or the governor.[78]

If these do not work, further tactics of psychological violence are implemented by taking the student under detention. As the police conduct surveillance for long periods, they use a student's activity as a pretext to make an arrest.[79] The creation of a "spectacular" scene with a police raid is the first step. The police usually initiate the process early in the morning at 5 a.m. and arrive at the neighborhoods fully armed in armored vehicles with special operation teams and helicopters. Students are taken to Anti-Terrorism Branches. One of the tactics they use during detention is to interrogate after midnight. Student A explained it in this way: "They ruin your sleep pattern and wake you up every 3 hours ... They inflict psychological torture ... You cannot think healthy." They show them surveillance records; they imply that they know every detail of their lives and threaten them with long prison terms.

Preemptive risk-elimination techniques culminate in the initiation of a court case to neutralize ungovernable students/risks. The ongoing Istanbul Case of Democratic Patriotic Youth filed in 2011 is a good example, since 72 students have been prosecuted under it, with little in the way of evidence. Most accusations are based on non-violent political activities. 31 of these students were put in prison on remand to be gradually released. Some students were kept in prison for more than two years. The indictment includes, as usual, a highly political narrative about the Democratic Patriotic Youth Assemblies. These assemblies formed the youth organization officially affiliated with the Peace and Democracy Party, the political party of the Kurdish movement until 2014. However, in the indictment, it is represented as the youth organization of the PKK by relying on confessions taken from some defendants in return of leniency in punishment. The main arguments of the indictment are about two meetings held in January 2011 at the Istanbul office of the party. It is argued that in these meetings decisions on the PKK activities for protesting the anniversary of its leader's capture were taken. The activities included car burning and throwing molotov cocktails. According to the indictment, most of the defendants attended these meetings. They are accused of being members of a terrorist organization and making its propaganda. As with most defendants, there is no evidence proving that they have attended these meetings or these activities. Confessions are frequently used to incriminate them mostly for their non-violent political activities such as participating in demonstrations. Hence, these court cases serve as important instruments in the pacification of the Kurdish students and the Kurdish liberation movement at large.[80]

CONCLUSION

Student activities taking place at university campuses in Turkey as part of the Kurdish liberation movement reveal the exploitative/authoritarian nature of state strategies on the Kurdish issue, which offer nearly no improvement. By examining the mechanisms through which these students are policed, one can arrive at some conclusions. First of all, as this student movement that challenges the existing political-economic order has been criminalized with the activation of various "security measures," the campuses are spaces in which the pacification of social opposition and depoliticization of the socio-political field for the ordering of social relations of power[81] take place. Furthermore, what is called "security" is an important mediation point for the capitalist state to incessantly reconstruct order at various spatial levels. University campuses lay bare how these security measures materialize on the ground.

A further conclusion is that historicizing the strategies employed within the pacification process reveals their novel workings. In the paper, the newly established surveillance-policing networks at campuses in Turkey in late 2000s and their functioning are analyzed as part of the novel pacification mechanisms of the 'security state' emerging within the neoliberalization process. As this state form rests on neoliberal governmental rationality and prioritizes the prevention of 'risks' to achieve cost-reduction and efficiency, intelligence gathering acquired further significance and the penal sphere is infiltrated with logic of "preemptive security." In the spaces of the higher education that are reconfigured as "durably pacified social spaces"[82] this led to the establishment of surveillance-policing networks to extensively monitor the campuses, inflict psychological violence for repressive and disciplinary purposes, and implement preemptive "risk" elimination techniques. As in the case of the Kurdish students, the state tries to pacify alternative political stances, subsume them into the existing relations of exploitation, and "remodel expectations about political rights, individual liberties and social freedoms, all in the name of security."[83]

ACKNOWLEDGMENTS

I am grateful to Assoc. Prof. Zeynep Kıvılcım for the precious help she provided me in holding the interviews and preparing the article. I also owe many thanks to Assist. Prof. Doğan Çetinkaya, Bahadır Ahıska and the students for sharing their tough experiences they had with the state at various levels.

NOTES

1 Dylan Rodriguez, "Beyond 'Police Brutality': Racist State Violence and the University of California," *American Quarterly* 64, no. 2 (2012): 301-313, 311.
2 Henri Lefebvre, "Space and the State," in *State/Space-A Reader*, eds. Neil Brenner, Bob Jessop, Martin Jones, and Gordon Macleod (Malden: Blackwell Publishing, 2003), 84-100, 84-86.
3 Piya Chatterjee and Sunaina Maira, eds., *The Imperial University: Academic Repression and Scholarly Dissent* (Minneapolis: University of Minnesota Press, 2014); Anthony J. Nocella II, and David Gabbard, eds., *Policing the Campus-Academic Repression, Surveillance and the Occupy Movement* (New York: Peter Lang Publishing, 2013); Anthony J. Nocella II, Steven Best, and Peter McLaren, eds., *Academic Repression-Reflections from The Academic Industrial Complex* (Oakland: AK Press, 2010).
4 See Nicos Poulantzas, *Classes in Contemporary Capitalism* (London: Verso Edition, 1978).
5 Mark Neocleous, "Security as Pacification," in *Anti-Security*, eds. Mark Neocleous and George S. Rigakos, (Ottawa: Red Quill Books, 2011), 23-56, 42.
6 Neocleous, "Security as Pacification"; Will Jackson, "Securitisation as Depoliticisation: Depoliticisation as Pacification," *Socialist Studies* 9, no. 2 (2013): 146-166; Tia Dafnos, "Pacification and Indigenous Struggles in Canada," *Socialist Studies* 9, no. 2 (2013): 57-77. Also see Will Jackson, Helen Monk and Joanna Gilmore's chapter in this volume.
7 "Preemption" is used here not to indicate to the prevention of a *concrete* danger. The implication in preemption is "intervention in advance," to prevent even the possibility of an unwanted incident leading to the interrogation, arrest and exclusion of the accused in advance. Susanne Krasmann, "The Enemy on the Border: Critique of a Programme in favor of a Preventive State," *Punishment and Society* 9, no. 3 (2007): 301-318, 307.
8 For an overview see Panagiotis Sotiris, "Theorizing the Entrepreneurial University: Open Questions and Possible Answers," *Journal of Critical Education Policy Studies* 10, no. 1 (2012): 112-126, 113-114.
9 See endnote 3.

10 Farah Godrej, "Neoliberalism, Militarization, and the Price of Dissent-Policing Protest at the University of California," in *The Imperial University*, eds. Chatterjee and Maira, 125-143, 127.
11 Sunaina Maira and Julie Sze, "Dispatches from Pepper Spray University: Privatization, Repression, and Revolts," *American Quarterly* 64, no. 2 (2012): 315-330.
12 Richard van Heertum, "Cameras and ID Card Swipes: Disappearing Privacy and the Cultivation of the Virtual Self," in *Policing the Campus*, eds. Nocella and Gabbard, 67-78.
13 Maira and Sze, "Dispactches from Pepper Spray," 322.
14 Peter Hopkins, "Towards Critical Geographies of the University Campus: Understanding the Contested Experiences of Muslim Students," *Transactions of the Institute of British Geographers* 36, no. 1 (2011): 157–169, 167.
15 Chris Philo and Hester Parr, "Institutional Geographies: Introductory Remarks," *Geoforum* 31 (2000): 513-21; cited in Peter Hopkins, "Towards Critical Geographies of the University Campus," 157.
16 Lefebvre, "Space and the State," 84.
17 Hopkins, "Towards Critical Geographies of the University Campus," 158.
18 Neocleous, "Security as Pacification," 35, 36.
19 Ibid., 24, 26.
20 See Neil Brenner, Jamie Peck and Nik Theodore, "Variegated Neoliberalization: Geographies, Modalities, Pathways," *Global Networks* 10, no. 2 (2010): 182-222.
21 Sotiris, "Theorizing the Entrepreneurial University," 113-116. See Henry A. Giroux, *The University in Chains: Confronting the Military-Industrial-Academic Complex* (Boulder: Paradigm Publishers, 2007).
22 Andrew McGettigan, " 'New Providers'—The Creation of a Market in Higher Education," *Radical Philosophy* 167 (2011): 2-8; David J. Blacker, *The Falling Rate of Learning and the Neoliberal Endgame* (Winchester: Zero Books, 2013).
23 Jason Del Gandio, "Arrests and Repression as a Logic of Neoliberalism," in *Policing the Campus*, eds. Nocella and Gabbard, 3-14, 10.
24 Jamie Peck, "Geography and Public Policy: Mapping the Penal State," *Progress in Human Geography* 27, no. 2 (2003): 222–232; Loic Wacquant, "The Penalisation of Poverty and the Rise of Neo-Liberalism," *European Journal on Criminal Policy and Research* 9 (2001): 401-412.
25 Poulantzas, *State, Power, Socialism*, (London: Verso, 2000), 159.
26 Simon Hallsworth and John Lea, "Reconstructing Leviathan: Emerging Contours of the Security State," *Theoretical Criminology* 15, no. 2 (2011): 141-157; Christopher Murphy, "'Securitizing' Canadian Policing: A New Policing Paradigm For the Post 9/11 Security State?" *Canadian Journal of Sociology* 32, no. 4 (2007): 449-475.
27 Poulantzas, *State, Power, Socialism*, 227.
28 Jean-Claude Paye, *Global War on Liberty* (New York: Telos Press Publishing, 2007), 185.
29 Poulantzas, *State, Power, Socialism*, 203-204.
30 The concept of "risk" here is based on the Foucauldian analyses taking it as a "way of representing events in a certain form so they might be made governable ... with particular techniques and for particular goals" rather than Ulrich Beck's conceptualization as a feature of the ontological condition of humans. Mitchell Dean, *Governmentality: Power and Rule in Modern Society* (London: Sage Publications, 2001), 177-178. As Castel puts forward, since late 1970s, "[a] risk does not arise from the presence of particular precise danger embodied in a concrete individual or group. It is the effect of a combination of abstract *factors* which render ... probable the occurrence of undesirable modes of behavior." Robert Castel, "From Dangerousness to Risk," in *The Foucault Effect: Studies in Governmentality*, eds. Graham Burchell et al. (Chicago: The University of Chicago Press, 1991), 281-298, 287. It leads to *preventive* policies and a new mode of surveillance for systematic *pre-detection*, Castel, "From Dangerousness to Risk," 288. Risk networks are constituted by agencies and institutions engaging in risk-profiling practices., Pat O'Malley, "Review Article: Governmentality and the Risk Society," *Economy and Society* 28, no. 1 (1999): 138-148, 141.
31 Their foundation is in the liberal art of government. See Michel Foucault, *Security, Territory, Population,* (New York: Palgrave MacMillan, 2007), 1-86.
32 Dean, *Governmentality,* 170.
33 Richard V. Ericson, "The State of Preemption: Managing Terrorism Risk through Counter Law," in *Risk and the War on Terror,* eds. Louise Amoore and Marieka de Goede (Oxford: Routledge, 2008), 57-76, 66.
34 Michael Dillon, "Underwriting Security," *Security Dialogue* 39 (2008): 309-332, 325.

35 Hallsworth and Lea, "Reconstructing Leviathan."
36 See endnote 8.
37 As Seri explains, Foucault defines the concept of *dispositif* as the combination of discourses, policies, institutions, architectural forms, laws, regulatory decisions, philosophic or moral propositions etc. to constitute a governing net based on a certain rationality to solve social problems. See Guillermina Seri, *Seguridad: Crime, Police Power and Democracy in Argentina* (New York: Continuum International Publishing, 2012), 4; Michel Foucault, "The Confession of the Flesh," in *Power/Knowledge: Selected Interviews and Other Writings. 1972-77*, ed. Colin Gordon (New York: Pantheon, 1980), 194-228, 194.
38 Ericson, "The State of Preemption," 57.
39 According to Hallsworth and Lea: "The security state searches for *new technologies of ... risk management* aimed at 'external' threats that, in a globalized world, *may originate in the next street* or in another continent.... [The whole population] becomes subject to *intensifying levels of surveillance and control in a process of 'securitization' that becomes the pre-eminent model for the regulation* of post-disciplinary 'societies of control'." "Reconstructing Leviathan," 142 (*italics added*).
40 Klaus Mladek, "Exception Rules: Contemporary Political Theory and the Police," in *Police Forces-A Cultural History of an Institution*, ed. Klaus Mladek (New York: Palgrave Macmillan, 2007), 221-265, 249.
41 Paye, *Global War*, 81, 107.
42 Claudia Aradau and Rens Van Munster, "Governing Terrorism Through Risk: Taking Precautions, (un)Knowing the Future," *European Journal of International Relations* 13, no. 1 (2008): 89-115, 103, 106.
43 Jackson, "Securitisation as Depoliticisation," 155. See also Jeffrey Monaghan and Kevin Walby, "Making up 'Terror Identities': Security Intelligence, Canada's Integrated Threat Assessment Centre and Social Movement Suppression," *Policing and Society* 22, no. 2 (2012): 133-151.
44 Poulantzas, *State, Power, Socialism*, 77, 83.
45 Michel Foucault, *History of Sexuality, Vol. 1* (London: Penguin Books, 1998), 133-160; Francois Ewald, "Norms, Discipline and the Law," *Representations*, 30 (1990): 138-161.
46 The main factors enabling this transformation are the changing composition of the power bloc in which the conservative bourgeoisie became a rival of the hegemonic Istanbul-based "secular" big bourgeoisie, the dissolution of the alliance between military and big bourgeoisie as well as shifting international conjuncture which became unfavorable for military tutelage. İsmet Akça, "Hegemonic Projects in Post-1980 Turkey and the Changing Forms of Authoritarianism," in *Turkey Reframed-Constituting Neoliberal Hegemony*, eds. İsmet Akça, Ahmet Bekmen and Barış Alp Özden (London: Pluto Press, 2014), 13-46, 31-38. Biriz Berksoy, " 'Güvenlik Devleti'nin Ortaya Çıkışı, 'Güvenlik' Eksenli Yönetim Tekniğinin Polis Teşkilatındaki Tezahürleri ve Süreklileşen 'Olağanüstü Hal': AKP'nin Polis Politikaları," ("The Emergence of the 'Security State,' Manifestation of the 'Security' Oriented Governmental Technique in the Police Organization and the Permanence of the 'State of Emergency': The Police Policies of the Justice and Development Party"), *Birikim* 276 (2012): 75-88.
47 Berksoy, "'Güvenlik Devleti'nin Ortaya Çıkışı," 75-88.
48 Akça, "Hegemonic Projects in Post-1980 Turkey," 30-44.
49 EGM (2008), "Stratejik Plan 2009-2013," http://www.egm.gov.tr/documents/emniyet_genel_mudurlugu_stratejik_plani.pdf, (Accessed: 21.12.2012).
50 Berksoy, "'Güvenlik Devleti'nin Ortaya Çıkışı," 82-86.
51 Ibid.
52 The transformation within the Kurdish liberation movement to incorporate masses of Kurdish people engaging in civil disobedience acts in 2000s has played a significant role in the crystallization of this rationality within the legal sphere. Derya Bayır, "The Role of the Judicial System in the Politicide of the Kurdish Opposition," in *The Kurdish Question in Turkey: New Perspectives on Violence, Representation and Reconciliation*, eds. Cengiz Güneş and Welat Zeydanlıoğlu (London: Routledge, 2014), 21-46.
53 If a person engages in activities significant enough to exert pressure on government and is in a leading position and/or if even a remote link can be established between that person and an "illegal" organization, then that person is evaluated as posing a risk/danger to the state. Human Rights Watch, *Protesting as a Terrorist Offense*, 2010, http://www.hrw.org/sites/default/files/reports/turkey1110webwcover.pdf, accessed March 6, 2014; İsmail Saymaz, *Sözde Terörist (So-Called Terrorist)* (İstanbul: İletişim Yay, 2013).

54 Ayşen Uysal, "Devletin Güvenliği ve Toplumsal Muhalefet Eylemleri: 'Kalemli Çete' Örneği," ("State Security and Acts of Social Opposition: The Case of 'Gang with Pens' "), *Birikim* 146 (2001): 64-84, 71, 72.
55 Eğitim-Sen Yüksek Öğretim Bürosu, *Yükseköğretimin Yeniden Yapılandırılmasına İlişkin Görüş ve Önerilerimiz*, (*Our Opinion and Suggestions Regarding the Restructuring of Higher Education*), (Ankara: Eğitim-Sen Yayınları, 2012).
56 Cansu Kılınçarslan, "Polisliğin Dönüşümü ve Genişlemesi: Üniversitelerde Polis Projeleri" ("The Transformation and Expansion of Policing: Police Projects at Universities"), (unpublished paper).
57 Mark Neocleous, *Critique of Security* (Montreal: McGill-Queen's University Press, 2008), 12.
58 Neocleous, *Critique of Security*, 5.
59 Birgün, "Üniversiteler Karakol Rektörler Komiser" ("Universities are Police Stations, Rectors are Police Chiefs"), October 8, 2010. The meetings would include university rectors, members of the university administrations, members of the police organization, representatives from the mayor's office and other related state institutions.
60 Interviews with two leftist former students politically active in mid-1990s (see endnote 70). Semih Tatlıcan, "1980 Sonrası Öğrenci Dernekleri," ("Post-1980 Student Associations"), *Birikim*, 73 (1995): 72-75.
61 Judicial statistics are not kept according to occupational status; it is not possible to give annual numbers of students on trial. The Initiative for Solidarity with the Students in Prison announced that there were around 800 university students in prison on remand in 2012 by relying on indictments, letters from students, etc. See the Initiative's report: Tutuklu Öğrencilerle Dayanışma İnisiyatifi, *Tutuklu Öğrenciler Raporu* (*Report on Students in Prison on Remand*), 2012, http://www.google.com.tr/url?sa=t&rct=j&q=&esrc=s&-source=web&cd=1&ved=0CBsQFjAA&url=http%3A%2F%2Fwww.bianet.org%2Ffiles%2F-doc_files%2F000%2F000%2F624%2Foriginal%2FT%25C3%2596D%25C4%25B0_tutuklu_%25C3%25B6%25C4%259Frenciler_raporu.docx&ei=PEWpVlfQJonV7Qaz3YD4Dg&usg=AFQjCNEfDY9uk6ZVL3Lzl0NCNsdMwA-PiHw, accessed August 27, 2012).
62 This is an estimation made by a member of the initiative. See the news titled "Türkiye'de yüzlerce öğrenci tutuklu yargılanıyor," ("In Turkey there are hundreds of students are in prison on remand"), December 2, 2012, http://www.bbc.co.uk/turkce/haberler/2012/12/121202_turkey_students.shtml, accessed August 15, 2014.
63 İrfan Aktan, "The AKP's Three-Faceted Kurdish Policy: Tenders for the Rich, Alms for the Poor, Bombs for the Opposition," in *Turkey Reframed*, eds. Akça, Bekmen and Özden, 107-121, 117.
64 Barış Ünlü, "Kürdistan/Türkiye ve Cezayir/Fransa: Sömürge Yöntemleri, Şiddet ve Entellektüeller," (Kurdistan/Turkey and Algeria/France: Colonial Methods, Violence and Intellectuals), in *Türkiye'de Siyasal Şiddetin Boyutları*, eds. Güney Çeğin and İbrahim Şirin (İstanbul: İletişim Yay., 2014), 369-401.
65 Zülküf Aydın, *Underdevelopment and Rural Structures in South-eastern Turkey: The Household Economy in Gisgis and Kalhana* (London: Ithaca Press, 1986).
66 Mesut Yeğen, *Müstakbel Türk'ten Sözde Vatandaşa: Cumhuriyet ve Kürtler* (*From Prospective Turk to So-Called Citizen: Republic and the Kurds*) (İstanbul: İletişim Yay, 2006), 9-45.
67 Ahmet Hamdi Akkaya and Joost Jongerden, "Confederalism and Autonomy in Turkey: The Kurdistan Workers' Party and the Reinvention of Democracy," in *The Kurdish Question in Turkey*, eds. Güneş and Zeydanlıoğlu, 186-204.
68 Akkaya and Jongerden, "Confederalism and Autonomy in Turkey."
69 Cengiz Güneş, "Political Reconciliation in Turkey: Challenges and Prospects," in *The Kurdish Question in Turkey*, eds. Güneş and Zeydanlıoğlu, 258-281.
70 Akça, "Hegemonic Projects in Post-1980 Turkey," 43.
71 Gökçer Tahincioğlu and Kemal Göktaş, *"Bu öğrencilere bu işi mi öğrettiler?": Öğrenci Muhalefeti ve Baskılar ("Did they teach this to these students?": Student Opposition and Repression)* (İstanbul: İletişim Yay, 2013).
72 Five Kurdish students study at Istanbul University; two students study at Mimar Sinan University of Fine Arts and Kocaeli University. Of the seven private security guards, three of them work at Istanbul University; two of them work at Marmara University; two of them work at Yıldız Technical University. Two former students, Doğan Çetinkaya and Bahadır Ahıska, were active within the leftist student collective called "the Coordination," the most influential student collective in mid-1990s.

73 The political orientation of the university rectors and administrations is important as to the degree these networks can realize their potential. Few universities, such as Boğaziçi University, are administered with a relatively non-interventionist attitude. However, as the rectors are appointed by the President of the Republic and the "Free and Secure University Project" is intended for all universities, it can be stated that they remain exceptional.
74 See the news titled "Kızınız terör örgütünün eyleminde" ("Your daughter is taking part in the activity of the terrorist organization"), July 31, 2012, http://www.bianet.org/bianet/ifade-ozgurlugu/140033-kiziniz-teror-orgutunun-eyleminde, accessed September 28, 2014).
75 Krasmann, "The Enemy on the Border," 307.
76 The police officers are mainly from three branches: branches of Security, Anti-Terrorism and Intelligence.
77 The number of investigations has risen from around 2600 in 2000 to around 6000 in 2011. See Benan Molu, Esra Demir Gürsel, Gülşah Kurt, Hülya Dinçer, and Zeynep Kıvılcım, Üniversitelerde Disiplin Soruşturmaları: Öğrencilerin İfade ve Örgütlenme Özgürlüğü, (*Disciplinary Investigations at Universities: Students' Freedom of Expression and Assembly*) (İstanbul: İki Levha Yayıncılık, 2013), 1.
78 Molu et al., Üniversitelerde Disiplin Soruşturmaları, 85-94.
79 Student C explained this process in this way: "The content of that declaration is not the reason for my arrest for 9 months ... I was active and I was engaging in politics ... I was one of the students that they most wanted to have arrested."
80 Istanbul Office of the Chief Prosecutor; Investigation Number: 2010/2483; Indictment Number: 2011/296; Tutuklu Öğrencilerle Dayanışma İnisiyatifi, *Tutuklu Öğrenciler Raporu*; Tutuklu Öğrencilerle Dayanışma İnisiyatifi, İkinci Tutuklu Öğrenciler Raporu (Second Report on Students in Prison on Remand) (unpublished report).
81 Neocleous, "Security as Pacification," 42.
82 Norbert Elias, *The Germans: Power Struggles and the Development of Habitus in the Nineteenth and Twentieth Centuries* (New York: Columbia University Press, 1996), 174, 176; cited in Neocleous, "Security as Pacification," 42.
83 Neocleous, "Security as Pacification," 48.

CPSIA information can be obtained
at www.ICGtesting.com
Printed in the USA
LVOW12s20533111017
554465LV00001B/115/P